Felice Carlo Simeone

Local characterization and modification of surfaces by the in-situ STM

Felice Carlo Simeone

Local characterization and modification of surfaces by the in-situ STM

from nanostructuring to local analysis

Südwestdeutscher Verlag für Hochschulschriften

Impressum/Imprint (nur für Deutschland/ only for Germany)
Bibliografische Information der Deutschen Nationalbibliothek: Die Deutsche Nationalbibliothek verzeichnet diese Publikation in der Deutschen Nationalbibliografie; detaillierte bibliografische Daten sind im Internet über http://dnb.d-nb.de abrufbar.
 Alle in diesem Buch genannten Marken und Produktnamen unterliegen warenzeichen-, marken- oder patentrechtlichem Schutz bzw. sind Warenzeichen oder eingetragene Warenzeichen der jeweiligen Inhaber. Die Wiedergabe von Marken, Produktnamen, Gebrauchsnamen, Handelsnamen, Warenbezeichnungen u.s.w. in diesem Werk berechtigt auch ohne besondere Kennzeichnung nicht zu der Annahme, dass solche Namen im Sinne der Warenzeichen- und Markenschutzgesetzgebung als frei zu betrachten wären und daher von jedermann benutzt werden dürften.

Verlag: Südwestdeutscher Verlag für Hochschulschriften Aktiengesellschaft & Co. KG
Dudweiler Landstr. 99, 66123 Saarbrücken, Deutschland
Telefon +49 681 37 20 271-1, Telefax +49 681 37 20 271-0
Email: info@svh-verlag.de
Zugl.: Ulm, University, Diss., 2008

Herstellung in Deutschland:
Schaltungsdienst Lange o.H.G., Berlin
Books on Demand GmbH, Norderstedt
Reha GmbH, Saarbrücken
Amazon Distribution GmbH, Leipzig
ISBN: 978-3-8381-0545-1

Imprint (only for USA, GB)
Bibliographic information published by the Deutsche Nationalbibliothek: The Deutsche Nationalbibliothek lists this publication in the Deutsche Nationalbibliografie; detailed bibliographic data are available in the Internet at http://dnb.d-nb.de.
 Any brand names and product names mentioned in this book are subject to trademark, brand or patent protection and are trademarks or registered trademarks of their respective holders. The use of brand names, product names, common names, trade names, product descriptions etc. even without a particular marking in this works is in no way to be construed to mean that such names may be regarded as unrestricted in respect of trademark and brand protection legislation and could thus be used by anyone.

Publisher: Südwestdeutscher Verlag für Hochschulschriften Aktiengesellschaft & Co. KG
Dudweiler Landstr. 99, 66123 Saarbrücken, Germany
Phone +49 681 37 20 271-1, Fax +49 681 37 20 271-0
Email: info@svh-verlag.de

Printed in the U.S.A.
Printed in the U.K. by (see last page)
ISBN: 978-3-8381-0545-1

Copyright © 2010 by the author and Südwestdeutscher Verlag für Hochschulschriften Aktiengesellschaft & Co. KG and licensors
All rights reserved. Saarbrücken 2010

Io pensavo ad un'altra morale, piú terrena e concreta, e credo che ogni chimico militante la potrá confermare: che occorre diffidare del quasi uguale (il sodio é quasi uguale al potassio), del praticamente identico, del pressappoco, dell'oppure, di tutti i surrogati e di tutti i rappezzi. Le diffrenze possono essere piccole, come gli aghi degli scambi; il mestiere del chimico consiste in buona parte nel guardarsi da queste differenze, nel conoscerle da vicino, nel prevederne gli effetti. Non solo il mestiere del chimico.

Primo Levi

We see that science and peace are related. The world has been greatly changed, especially during the last century, by the discoveries of scientists. Our increased knowledge now provides the possibility of eliminating poverty and starvation, of decreasing significantly the suffering caused by disease, of using the resources of the world effectively for the benefit of humanity. But the greatest of all the changes has been in the nature of war, the several million fold increase in the power of explosives and corresponding changes in methods of delivery of bombs.

These changes have resulted from the discoveries of scientists, and during the last two decades scientists have taken a leading part in bringing them to the attention of their fellow human beings and in urging that vigorous action be taken to prevent the use of the new weapons and to abolish war from the world.

Linus Pauling

Questo lavoro di tesi é dedicato a tutti gli italiani emigrati in Germania e quindi a tutti gli stranieri, legali o illegali, dispersi in ogni angolo del mondo.

This thesis is dedicated to all Italians emigrated to Germany and then to all legal and illegal foreigners of every corner of the world.

Diese Doktorarbeit ist allen Italienern gewidmet, die nach Deutschland ausgewandert sind, und allen legalen und illegalen Ausländern von jeder Ecke der Welt.

Contents

Introduction: nano-technology vs. nano-science   vii

1 The electrified interface   1
    1.1 The metal side of the interface 2
    1.2 The water/electrode interface 3
    1.3 The electrochemical *double-layer* 7

2 Tunneling in electrochemistry   9
    2.1 The electron tunneling 9
    2.2 Electron transfer at the electrified interface 14
    2.3 The Scanning Tunneling Microscopy (STM) 16

3 Experimental   23
    3.1 STM 23
    3.2 Electrodes 24
       3.2.1   Au(111) 24
       3.2.2   Ir(210) 24
    3.3 Nanostructuring 25
    3.4 Tunneling Spectroscopy 26
    3.5 Chemicals 27

4 Nano-faceting of surfaces: the case of Ir(210)   29
    4.1 Introduction 29
    4.2 Morphology and faceting mechanism 30
    4.3 Electrochemical behaviour of the faceted Ir(210) 42

5 Electrochemical nanostructuring with the STM   47
    5.1 Introduction 47
    5.2 Electrochemical tip-driven nanostructuring 50
       5.2.1   What is a *jump-to-contact*? 50
       5.2.2   Nanostructuring based on the Jump-to-Contact 52

CONTENTS

    5.3 Pd clusters on Au(111) 62
        5.3.1   Electrodeposition of Pd: an overview 62
        5.3.2   Pd clusters on Au(111) 63
        5.3.3   The stability of the clusters 66
    5.4 First investigation of Pd cluster reactivity with the SECM .. 73

6 Local Analysis of the Electrified Interface   83
    6.1 Introduction 83
    6.2 Some particular aspect of the *in-situ* STM 84
        6.2.1   The problem of the absolute gap width 85
    6.3 Investigation of the Au(111)/H_2SO_4 interface 87
    6.4 An electrochemical nano-switch 101
        6.4.1   Electronic properties of the interface 108
    6.5 Spectroscopic investigation of Pd thin films 112
        6.5.1   Structure of the interface with the DTS 112
        6.5.2   Voltage Spectroscopy investigation 118
    6.6 First investigations of Pd nano-clusters 120

7 Summary   125

8 Zusammenfassung   127

Introduction: nano-technology vs. nano-science

When I moved to Germany for my Ph.D, I was not completely aware of what I was going to do. All I knew was that I had to work on a project about nano-structures in electrochemistry. That was all: when my friends asked about my future job, I answered that I had to do nano-things!

At the beginning, a lot of my effort went into learning to work with an STM and in what kind of informations can be obtained with this technique.

After almost one year hanging around in the STM room and just some seconds before I could definitively conclude that the STM simply doesn't work, I got my first STM image with a clear and wonderful atomic resolution. That was my first fundamental result: atoms really exist and I can see them!

However, while I was daily fighting with the STM, I began to understand something more about my project. Indeed, none-strictly scientific reports on something called nano-technology started to appear almost everywhere: popular magazines, news-papers. Radio and TV shows were dedicated to the topic. It is not usual that a specific scientific subject reaches a broad audience as for the case of nano-technology.

I quickly realized the reason for such an attention: the laptop where I'm writing this thesis was almost new at the beginning of my work. After five years, laptops of the same size of mine have an incredibly higher data storage capability although they are cheaper and lighter.

In few years, the miniaturization of the components inside devices that we use every day is already deeply influencing our life.

Hence, I've begun to understand the reason why a lot of people, most of them not directly involved in science, pay so much attention to nano-technology.

At scientific meetings, very often the results discussed dealt with strategies for miniaturization of the matter down to the nano scale and it is really impressing the high degree of interdisciplinary character of this emerging field. Nano-technology is a term employed in physics as well as in medicine,

CHAPTER 0. Introduction: nano-technology vs. nano-science

by chemists as well as by biologists.

I was, and I am still impressed by the impact of the STM on the discovery of such a nano-world.

An idea of how important it is to be involved in nano-technology, can be given by the fact that very often the prefix *nano* is an abused word. Indeed, in order to attract the attention of an audience or just to get a project granted, the terms nano-technology and nano-science are used for the description at atomic level of a macroscopic sample: the essence of nano-technology is that the size of the entire sample does matter!

Many, if not all countries of the world are promoting and financing the research in nano-technology since it is evident that the possibility to develop and to own such kind of knowledge will play an important role for development of the global economy of the next years. And not only economy.

Few months after the beginning of my Ph.D., I was invited to attend a meeting where a company presented the latest instruments for working at the nano-scale. They tried to convince us to buy the instruments by stressing that they were developed for military use. Later I discovered that among the major investors in nano-technology there is the Pentagon, the operative head quarter of the defense department of the USA government. It is clear that nano-technology can have the opposite aim of improving the life of human beings: it can be used in order to better destroy them.

Scientists are determinant in this process, and sometimes I have the feeling that not all of them are aware that the choices they will make will strongly influence the life of billion of peoples. In my opinion, an *ethical* issue on their work should be arisen also by scientists. Actually, I did my Ph.D. in Einstein's birth place!

Anyway, these few considerations can already give an account for the increasing number of scientific articles on nano-technology on both general and new, specific journals where with nano-technology are meant all the processes and methods for a controlled production of nano-sized matter.

The description of what matter really is when reduced to the nano-scale and what it does, is the aim of nano-science, a little bit less popular than nano-technology.

Very often the present nano-era is regarded as a new revolution. What kind of revolution? What is changing?

Already Richard Feynman, who many consider the initiator of the nano-era, argued that there is nothing in the well known laws of the quantum mechanics which prevents from handling with matter at the atomic scale[1]. Actually, the emerging nano-science seems not to be based on new and un-

---

[1] Richard Feynman, *There's plenty of room at the bottom*, talk at Caltech, 1959

known laws of the nature. Its concepts and postulates have already been formulated.

Most likely we are in one of those periods following a scientific revolution when the new concepts are re-formulated, as claimed by Thomas Kuhn[2]. We are still *digesting* the earthquakes of the relativity and of the quantum theory.

What's new, then? What we usually call iron or gold, are materials which behave in a precise way. It can be daily experienced that they conduct electricity and heat, regardless if we are speaking about a small gold earring or a railway iron line extending for kilometers. They are metals.

But, by reducing the dimensions of the earring down to the atomic scale, it will appear really different of that of our girl friend.

In agreement with the quantum theory, when electrons are confined in a box with the size L of the same order of magnitude of their wavelength or smaller, only discrete energy values are allowed with a characteristic separation between the first two levels of:

$$\Delta E_{conf} = \frac{3h^2}{8mL^2} \qquad (1)$$

where m is mass of the electron. Eq. 1 doesn't depend on the material and is defined only by fundamental quantities and by L. This is one of the aspects which attract the attention of the scientists, namely the possibility to design systems with a controlled energy structure by choosing an appropriated L.

Just as an example, let consider the electron or thermal transport. When the system size L is in the range of the mean free path of the electrons, the transport mechanism switches from a diffusion-type to a ballistic one.

If it is quite easy to observe discrete energy levels at low temperatures, at room temperature the following relation must be fulfilled:

$$\Delta E_{conf} \geq kT \qquad (2)$$

It can be easily proven that the relation is satisfied for $L < 10 nm$.

Hence, it seems that the nano-earring conducts electricity and heat in a very strange way. I would also doubt whether it can still be used as jewel: it shouldn't shine that nice! All in all, one should refer to the nano-earring not anymore as to gold!

I guess that women are thankful for the classical behaviour of their earrings. And men too!

At the nano-scale, the reduced dimensions have other important consequences. It has been observed that nano-sized metal clusters have a different

---

[2]Thomas Kuhn, *The structure of scientific revolutions*, 1962, Chicago

CHAPTER 0.Introduction: nano-technology vs. nano-science

structure when compared to the macro-crystals made by the same atoms, and in any case the ratio between atoms at the surface and those in the bulk phase is extremely high for the clusters. All those aspects have a crucial influence on the behaviour of such nano-clusters.

We can speak about *iron* or *gold* as long as their size is much bigger than their electronic wavelength. Down to this level, the usual concepts of *gold* and *iron* for describing the behaviour of nano-sized bodies made by gold and iron atoms are not anymore appropriated: at the nano-scale, gold and iron are not anymore common metals.

I can't forget a question often put by my professor of statistical thermodynamics when I was a student: what if we were made only by 50 atoms? I can now answer that in that case the strange quantum-behaviour of the nano-earrings were absolutely normal and it would be very difficult to imagine a not quantized, classical world where the behaviour of earrings is better described (fortunately!) by the gravitation laws!

This is, in my opinion, the revolution: nano-science is making us aware of how important is the concept of *size* in the description of nature. But, at the same time, nano-science is just reminding us that we have a *relative* idea of the nature due to the fact that as human beings, we are made of many billions of atoms.

However, only when a scientific description of nature is able to radically change the perception we have of the reality and even influence the idea we have about our self, also as consequence of a renewed relationship with that reality, only in that case one could speak about a revolution.

Has nano-science already passed this test?

Is it enough, the broad availability of nano-technology based laptops for considering this revolution already concluded?

Ulm, January 2008

Chapter 1

The electrified interface

The interface is the region where two immiscible phases come in contact. The properties of such a phase boundary are caused by the rupture of the homogeneity and isotropy of the relative bulk phases and the asymmetric character of the systems rules its behaviour.

The aim of electrochemistry is the study of an interface which is held under the control of an external potential, the so called *electrified interface*.

That's not a surprise that through all the history of electrochemistry a lot of effort has gone into the knowledge and the detailed modeling of such a system, a work which can not yet be considered concluded. In chapter 6 of the present thesis a special application of the STM for the analysis of the metal/liquid interface, is just one more small contribution to this effort.

Although the liquid/liquid and solid/solid systems have gained in recent years an increasing attention of electrochemists [2, 3], the metal/liquid system remains the better understood and its properties are in the following outlined.

A convenient parameter, experimentally accessible for the study of the polarizability of the interface is its *capacity*. By recalling the notions of electrostatic, a parallel-plate condenser of unit area has a capacity given by:

$$C \equiv \frac{dq}{dV} = \frac{\epsilon}{4\pi d} \quad (1.1)$$

where d is the distance between the two plates. It is evident that the values assumed by Eq. 1.1 depend only on the geometry of the condenser.

In contrast with Eq. 1.1, the capacity of an electrified interface depends on the chemical and physical nature of the constituents and on the surface charge density. Due to its particular structure, the capacity C of an interface can be expressed in a formal way in terms of different contributions, namely:

$$\frac{1}{C} = \frac{1}{C_m} + \frac{1}{C_i} + \frac{1}{C_{sol}} \quad (1.2)$$

CHAPTER 1. The electrified interface

where C_m refers to the contribution of the metal, C_i is the internal capacity and accounts for the effect of the first layer of electrolyte in contact with the electrode surface, and C_{sol} for the contribution of the electrolyte not in direct contact with the electrode.

Although the properties of the electrified interface are the result of the interplay of these different factors which hence influence one each other, the distinction between *separate* contributions made in Eq. 1.2 is done in order to rationalize the behaviour of the system. The discussion of these single terms is the object of the following sections.

1.1 The metal side of the interface

Since electrochemistry has begun to develop also as surface science, the behaviour of the surface of a metal electrode was thought to be due to the influence of the liquid side of the interface.

In a pioneering work of Trasatti [1] the role of the nature of the metal in contact with an electrolyte was demonstrated and progressively more refined models for the metal surface of an electrode started to be developed.

Nowadays, it is accepted that the electronic properties of the metal independent of the presence of an electrolyte, are important in the understanding of several experimental results.

It became also clear that not only the electrical properties of the metal were important. Due to the very small distances between metal surface and the particles constituting the liquid phase, the quantum mechanical behaviour of the electrons populating the interface can't be anymore neglected.

It is well known that electrons can penetrate into classically energetically forbidden regions, as it will be extensively discussed in section 2.1.

At a metal surface, this special behaviour of sub-atomic particles results in a spill-over of the electron density and an excess of negative charge is smeared outout from the metal surface. This situation gives rise to a surface dipole with the positive pole pointing to the metal.

A similar behavior has been also observed parallel to the surface. Due to the anisotropy of the interactions at the surface, the electrons will preferentially occupy holes or valley where they are closer to the positive cores of the crystal. It's not hard to expect that this phenomenon, known as *Smoluchowski effect* [4], will be greater on rough surfaces, and for single crystals, at more open surface. This lateral redistribution of the electrons contrasts with the spill-over normal to the surface and is responsible for a contraction of the distance between the first layers of the crystal lattice. Experimental measurements of the work function on different single crystal faces account for this phenomenon. Indeed, at densely packed orientation, e.g. the (111) of a fcc, the spill-over of the electrons is smaller than that at a (110) and that causes a more reduced work function for the more open (110) surface [2].

Anticipating what will be discussed in the next sections, the molecules of the electrolyte will also influence the distribution of the charge density at the metal surface, enhancing or hindering both the spill-over and the Smoluchowski effect.

Another important geometrical modification of the surfaces is the reconstruction, a displacement of the atoms at the surface from their lattice sites as response to the anisotropy of the surface energy.

All these phenomena contribute to the change of the surface potential χ with the surface-charge density σ_m giving rise to the metal capacity C_m which can be defined as [5]:

$$\frac{1}{C_m} = \frac{\partial \chi}{\partial \sigma_m} \qquad (1.3)$$

The quantity expressed in Eq. 1.3 is not a constant related to the geometry, but is a consequence of the high polarizability of the metal surface: an increase of the capacity C_m with the electronic density of the metal, which is observed in experimental data, can be therefore attributed to a corresponding increase in the surface polarizability.

Once the importance of the metal capacity C_m has been recognized, theoretical electrochemists encountered great difficulties in developing models capable to reproduce the experimental evidences, because, as already said, the different contributions to the total capacity of the interface are not independent.

Anyway, a model which can reproduce trends and order of magnitude of polycrystalline electrodes is the so called *jellium* [6, 7]. In this model the lattice of positively charged metal ions is represented by a constant positive background charge which drops abruptly to zero at the surface. The electrons are modeled as an homogeneous electron gas which interacts with the positive charge. This configuration choice for the electrons allows to properly describe the polarizability of the surface and then the metal-side capacity C_m.

The capacity of the high ordered single crystal surfaces can be theoretically reproduced by introducing periodic potentials for the positive charge distribution[8].

1.2 The water/electrode interface

Either liquid or solid, the bulk phase of water can be considered as a network of molecules bound together through hydrogen bridges. In the liquid

CHAPTER 1.The electrified interface

phase, the H_2O molecules participate in two hydrogen bonds forming long strings. In the crystal form, an oxygen atom is tetrahedrically coordinated to four surrounding hydrogens. Although its strength is not comparable with a chemical bond, the hydrogen bridge determines the structure and most of the properties of water.

By immersing a metal electrode in water, the network of molecules gets broken in correspondence of the interface and an anisoptropy is introduced.

The balance of the water-surface and water-water interactions determines the local structure of the interface. From an electrochemical point of view, the knowledge of the behaviour of the water at the interface is important because a significant number of molecules will be always in contact with the electrode, influencing the properties of the entire interface.

Let us start by bringing the metal surface described in the previous section in contact with water, and we consider first the case of a not charged surface. In a first approximation the structure of the interfacial water is determined only by not electrostatic interactions with the surface. A strong affinity of the atoms of the surface for oxygen will orient the water molecules with their oxygen pointing to the surface [9]. For the uncharged case, this is usually not the case of metals. In fact, the water molecules will orient their dipole moment preferentially parallel to the surface. Anyway, the presence of electrical dipoles in direct contact with the surface influences the response of the interface to a charge sent to the electrode, in other words, its capacity.

This capacity due to the very first layer of electrolyte on the electrode surface is the inner layer capacity C_i in Eq. 1.1 which is usually experimentally measured in very diluted solutions.

In this case, in proximity of the potential of zero charge (pzc) the differential capacity of the interface reaches a maximum.

As the potential is displaced from the pzc, a net charge flows to the surface and the water dipoles re-orient in order to minimize the electric field, determining an high capacity for the interface.

Nevertheless, the maximum doesn't appear exactly at the pzc. An asymmetry in the differential capacity vs. potential curve appears with the expected maximum at a slightly positive potential [10]. Some other effect must play a role in determining the water-electrode capacity besides the orientation of the dipoles under the effect of the electric field.

Indeed, the water molecules in contact with the surface are not monomers. The orientation of the dipole moment parallel to the surface allows for the formation of an hydrogen bonds network covering all the surface. The positive shift of the maximum in the differential capacity at the water/metal interface can be explained by considering that in order to align the molecules with the electric field, an activation barrier arising from the rupture of the

1.2 The water/electrode interface

two-dimensional network at the *pzc* has to be overcome. Once the network is broken, the molecule is free to turn the dipole in response to the electric field and hence increasing the capacity of the interface.

Something more needs to be said regarding the water interaction with a not charged metal surface. In the previous section the spill over of the electrons from the metal lattice has been discussed. A water molecule positioned within the electron gas will experience a net positive charge on the metal side which tends to turn its dipole with the negative end to the surface already at the *pzc*. Furthermore, chemical interactions of the water molecule with the metal atoms play an inportant role for the orientation of the dipole at the *pzc*. Anyway, this preferential orientation introduces a further barrier in the rotational movement of the water molecule contributing to the shifting of the maximum in the differential capacity from the *pzc* [5].

It should be also mentioned, that the corrugation of the surface could induce a monomerization of the water molecules in direct contact with the surface. A (110) orientation shows grooves through all the surface where water molecules can be trapped, hindering the formation of the surface network [11].

At potentials different of the *pzc*, the electric field plays a major role in determining the structure of the water at the surface.

At positive potentials, the two hydrogens point in direction of the solution. In this configuration the water molecules are able to form four hydrogen bonds with the surrounding molecules, giving rise to the so called *Ice-like bilayer* structure [5, 11, 12]. In the ideal model of the bilayer, the H_2O molecules can form three-dimensional structures in which the first layer is bound directly to the surface sites of an hexagonal pattern, and the molecules in the second layer are held by two or three hydrogen bonds to the first layer. Experimental results proving the existence of the bilayer also at the electrified interface have been reported [13, 14].

At further positive potentials, the charge at the surface will cause the penetration of anions in the inner layer of the interface removing the water molecules, as will be explained in the next section.

At potentials negative of the *pzc*, the hydrogen bonds network resists to the electric field created by a small negative charge. At further negative charge, the hydrogen bonds are disrupted and the water molecules react to the electric field by turning the hydrogens to the surface.

It has been noted that the water molecules rotate their dipole by keeping the oxygen still in contact with the electrode [13, 14]. In this way, hydrogen bonds between neighboring molecules are no longer possible.

At more negative charges, experimental measurements of the distance between the surface and the H_2O-oxygen indicate that the dipole is almost

CHAPTER 1.The electrified interface

perpendicular to the surface and that the oxygen has left its position in direct contact with the metal [15–17]. These authors also reported areal densities for the first layer greater than ice or bulk liquid water, resembling those found around hydrated ions. It is also worth to notice, that the profile of the oxygen density reported in these studies, follows an oscillating decrease with the distance from the surface, indicating a layering of water molecules extending for distances of several water molecular diameters. It is very difficult for cations to penetrate into the interfacial region, that's why also at negative potential the water molecules build ordered structure by maximizing the number of hydrogen bonds.

On the theoretical side, Spohr performed molecular dynamics simulations of water in contact with different electrodes showing that under these conditions the electrode-near area of the electrolyte (6-8 Å) forms well ordered layer-like structures comparable with the hexagonal ice (Ice-Ih) [18].

The structure of this bilayer has been intensively discussed in the literature over the last years, including the orientation (H-up or H-down) as well as the dissociation of the perpendicular-oriented water molecules (ref. [19] and references therein).

Theoretical calculations show also that the presence of water in direct contact with the surface has a pronounced effect on the electronic distribution of the metal. The presence of a dipole stabilizes, and hence enhances the spillover of the electrons: the charge is not anymore clearly separated like in a condenser but is also diffused all around the molecules. Furthermore, the charge separation on the water molecules induces an image charge on the metal side with which it interacts [5, 20]. All these effects contribute to C_i.

Another important issue must be mentioned. Being immobilized in a stable network, translational, rotational and vibrational solvation modes of the water molecules are strongly hindered. Only the optical polarization can still fully react to an external charge. The dielectric constant of water has a minimum at the surface which gradually, layer after layer, reaches its bulk value. This latter point marks one more difference with a plates-condenser and is important for the understanding of some features of the electron transfer at the electrified interface, as will be discussed in chapter 6.

The contribution of water to the capacity of the interface is experimentally accessible only in extremely diluted solutions, and reaches values far too high compared with the electrode/electrolyte system. But in any case, also for not diluted solutions the structure and the behaviour of the water at the interface will always play a crucial role for the processes occurring at the interface.

1.3 The electrochemical *double-layer*

In this section, the structure of the interface between a metal electrode and an aqueous solution containing inorganic ions is discussed.

Let's start with an electrically neutral surface covered with the water-network described in the previous section. In the bulk of the solution, the electric field created by the ions *organizes* the surrounding water molecules in a stable structure and they move around under thermal drift without losing this solvation shell.

When the solvated ions approach the electrode, in case of an high affinity with the atoms of the surface, they can get in direct *contact* with the surface.

This contact adsorption mechanism is not driven by electrostatic forces, and, thermodynamically speaking, is the result in a net energy gain in the process of desolvating (totally or partially) the ion, desolvating the electrode by disrupting the water network and forming a chemical bond between ion and metal. The fact that contact adsorption occurs already at the *pzc* and, as it has been experimentally proven, involves precise sites of the surfaces, demonstrates the chemical nature of the interaction which is called *specific adsorption*.

Due to their bigger size and polarizability, anions have a lower energy of solvation in comparison with cations which, in turn, retain always their solvation shell with maybe the only exception of Cs^+, which is already a *big* ion. That's not a surprise that specific adsorption is observed prevalently for anions.

The contact adsorbed species tend to form ordered structures at the surface and in order to describe these patterns, the plane passing through the centers of the specifically adsorbed ions is defined as the *Inner Helmholtz Plane* (IHP).

An experimental increase in the capacity of the inner layer with the decrease of the atomic radius of the specifically adsorbed species has been observed, in analogy with the specific capacity of a two plates condenser of Eq. 1.1 which increases by bringing the plates closer together.

The properties of the specifically adsorbed ion layers have much in common with the behaviour of water in contact with the surface of an electrode. The issue of a diffuse charge distribution and of the image-charge interactions are fundamental for the description of the IHP, but the chemical nature of the interaction complicates the picture.

In a chemical, specific interaction the electrons are redistributed as consequence of the formation of a chemical bond. In this case the ions do not retain the entire charge of the isolated ion. This *partial charge transfer adsorption* is not so easy to determine exactly because it is the result of a very

CHAPTER 1.The electrified interface

complicated process of donation of electrons from the ion to the metal followed by a back-donation from the metal to the ion, rearranging the charge also in molecular orbitals which are empty in the isolated ion.

Furthermore, the dipoles created by the contact-adsorbed ions, interact with the water molecules which still remain on the surface and contribute to the character of the IHP. All this interplay of electrical and chemical interaction determines the total capacity C_H of the inner layer, where the subscript H refers to Helmholtz, and where the contribution of the metal side as been already included.

By charging the electrode more positively, more anions get specifically adsorbed till a maximum is reached. Due to the lateral repulsion, this highest coverage is usually lower than a complete monolayer. In this case the charge on the surface can't be anymore balanced only by specific adsorption. Moreover, it is not seldom that neutral species populate the IHP and also the absence of specific adsorption is frequent. In these cases the net charge on the metal has its counter part in the region of the *not-specific adsorption*.

The not-specific adsorption is driven only by electrostatic forces which attract the ions together with their solvation shell till the closest approach to the surface is reached: the *Outer Helmholtz Plane* (OHP).

The thermal drifts smear out the not-specific adsorbed ions giving rise to a diffuse layer starting from the OHP. This mechanism of compensation of the electrode charge was modeled by Gouy and Chapman and predicts an exponential decay of the electrical potential through the diffuse layer [10]. The variation of the diffuse charge with the applied potential determines the capacity C_{sol} in Eq. 1.2.

The distinction between a solid and liquid side is the reason why it is referred to the structure of the electrified interface as *double-layer*, in agreement with an original idea of Stern [21], successively developed by Grahame [22] and Parsons [23] and is the current model accepted for the structure of the electrified interface.

Chapter 2

Tunneling in electrochemistry

2.1 The electron tunneling

The tunneling is a transport mechanism in condensed matter that allows a particle obeying the laws of the quantum mechanics to cross a potential barrier also when its total energy is lower than the barrier height. It is evident that the phenomenon can't be explained within the classical physics where the tunneling is simply impossible!

Indeed, it is the wave-particle dualistic behaviour of subatomic particles which makes possible for an electron with total energy E to penetrate into a classically forbidden region of potential $V > E$, as sketched in fig. 2.1a.

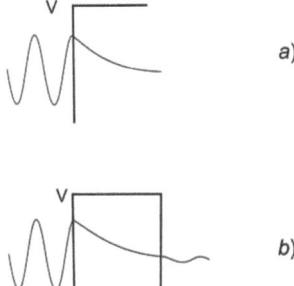

Figure 2.1: *Qualitative representation of a) a wave function extending into a classically forbidden region, and b) the tunneling through a potential barrier.*

When the potential barrier has a limited extension, the electron can *tunnel* from one side to the other (fig. 2.1b).

CHAPTER 2.Tunneling in electrochemistry

Since through all the present work it will be often referred to the parameters with which the tunneling process is described, its quantum-mechanical treatment is in following outlined.

Let's consider the system shown in fig. 2.2. An electron with wave vector

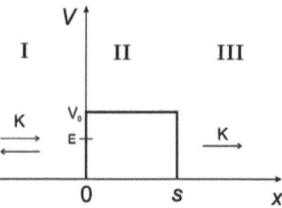

Figure 2.2: *One-dimensional potential barrier of height V and width s*

K and energy E moving from region I to region III has to cross three regions with different potential V:

- *for $x < 0$, $V = 0$* (region I in fig. 2.2)
- *for $0 < x < s$, $V = V_0$* (region II)
- *for $x > s$, $V = 0$* (region III)

The respective wavefunctions for the three regions must be solutions of the following Schrödinger equation:

$$\frac{d^2\Psi}{dx^2} + \frac{2m}{\hbar^2}(V-E)\Psi = 0 \qquad (2.1)$$

Standard solutions of Eq. 2.1 for the three regions are:

$$\Psi_I = Ae^{ikx} + Be^{-ikx}$$
$$\Psi_{II} = Ce^{ik'x} + De^{-ik'x} \qquad (2.2)$$
$$\Psi_{III} = Fe^{ikx} + Ge^{-ikx}$$

By substituting them in Eq. 2.1, the wave-vectors assume the following values:

$$k = \frac{\sqrt{2mE}}{\hbar} \qquad \text{(regions } I \text{ and } III\text{)}$$

$$k' = \frac{\sqrt{2m(E-V_0)}}{\hbar} \qquad \text{(with } V_0 > E \text{ region } II\text{)} \qquad (2.3)$$

2.1 The electron tunneling

Eq. 2.3 demonstrates that the Schrödinger equation admits a solution also in region II, where the wave-function has a real argument.

The probability that a particle crosses the barrier is expressed by the *transmission coefficient* T, defined as the ratio of the out going particle density F into the region III and the density A of the incident particle in region I:

$$T \equiv \left|\frac{F}{A}\right|^2 \qquad (2.4)$$

Analogously, a reflection coefficient R can be defined as $R \equiv B^2/A^2$ so that $R + T = 1$.

In order to evaluate T, the coefficients of the functions 2.2 are calculated by imposing the continuity of the function and of its first derivative at the borders between the different regions shown in fig 2.2.

Under these conditions, the transmission coefficient T for a particle with total energy E crossing a potential barrier with height V_0 and width s results:

$$T = 16\frac{E(V_0 - E)}{V_0^2} exp\left(-2s\frac{\sqrt{2m(V_0 - E)}}{\hbar}\right) \neq 0 \qquad (2.5)$$

Eq. 2.5 demonstrates that a particle has a *finite* probability to cross the forbidden area, probability which decreases exponentially with the increase of the barrier height and width.

A time dependent approach to the tunneling problem is very helpful for a better understanding of the process. The time-dependent Schrödinger equation is:

$$H\Psi(t) = i\hbar\frac{d\Psi(t)}{dt} \qquad (2.6)$$

with $H = (H_l + H_r) + H_T = H_0 + H_T$, where $H_l(H_r)$ is the Hamiltonian for the left (right) side of the tunneling junction, and H_T is the transfer Hamiltonian.

By substituting in Eq. 2.6 a general wave-function of the following form:

$$\Psi(t) = c(t)\Psi_l e^{-iE_l/\hbar} + d(t)\Psi_r e^{-iE_r/\hbar} \qquad (2.7)$$

one can identify the quantity:

$$M_{rl} = \int \Psi_r^*(H - E_l)\Psi_l dz = \int \Psi_r^* H_T \Psi_l \qquad (2.8)$$

with the effective tunneling matrix element. In general, the exact calculation of these matrix elements, when possible is a very complicated and time consuming task. Therefore, several strategies for approximated evaluations of these integrals have been proposed during the years.

CHAPTER 2. Tunneling in electrochemistry

Among others, Bardeen proposed to use the Fermi's golden rule of first order perturbation theory for strongly attenuating potential barriers [24]. With this approach the transmitted current j_t flowing through the tunneling junction assumes the following expression:

$$j_t = \frac{2\pi}{\hbar} |M_{rl}|^2 \frac{dN}{dE_r} \qquad (2.9)$$

The very important goal of Eq. 2.9 is the direct correlation of the tunneling current with density of the final state dN/dE_r.

Anyway, an exact calculation of the integrals in Eq. 2.8 yields a more complex expression for j_t, which results not anymore simply related to the density of states.

Besides the Bardeen's method, the so called WKB-approximation (Wentzel-Kramers-Brillouin) considers potential barriers of arbitrary shape and has been developed for practical applications.

In this model the probability $D(E)$ that an electron incident on a potential wall with total energy E can emerge on the opposite side of a potential barrier of width s and height $V = V(s)$ variable with the position s is done by:

$$D(E) = exp\left\{-\frac{2}{\hbar} \int_0^s [2m(V(s) - E)]^{1/2} ds\right\}$$
$$\equiv exp\left\{-2 \int_0^s \chi(s, E) ds\right\} \qquad (2.10)$$

The WKB approximation is adequate only if the energy E is not too close or above the top of the barrier. In any case, the WKB method has been very useful for the characterization of real tunneling junctions because allows for a more precise description of the barrier shape.

In all treatments presented above, the conservation of the energy has been assumed: that's what is called *elastic tunneling*. This is not always the case. The particle could indeed interact with elementary excitations where the energy is dissipated and not anymore conserved: this case refers to *inelastic tunneling*. It is known that the particle can be also trapped in a potential well giving rise to interference phenomena which delay the emerging of the particle on the opposite side of the gap. This is what is called a *resonating tunneling*. These different types of tunneling processes have been mentioned only for the sake of completeness, and their treatment goes beyond the aim of this work.

The heart of the present thesis is the electron elastic tunneling through a metal/insulator/metal junction. The discussion of some important feature of this system is useful for a better understanding of the results which will be later presented.

2.1 The electron tunneling

Let's start considering just one side of the tunnel junction. What does it mean for an electron to leave the metal surface? In surface science a *work function* Φ is defined as the work made on the electron for removing it from the solid's Fermi level outside an atomically flat and infinitely extended surface to the far vacuum level where the electron is free.

Real solids are far from being atomically flat and indefinitely extended, that's why a real an experimentally accessible *total work function* has been introduced.

Keeping in mind the different meanings of the two quantities, no distinction is anymore made between them in the following.

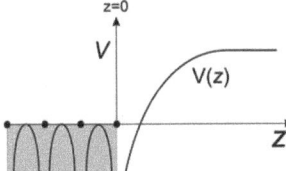

Figure 2.3: *One dimensional potential profile inside a metal lattice* $(z < 0)$, *and outside the surface* $(z > 0)$

As it can be seen in fig. 2.3, the potential outside a metal surface doesn't reach immediately the value of the vacuum level because at small distances the electron experiences the presence of the surface through interactions which can be formally distinguished between (see section 1.1):

- chemical forces (exchange, correlation)

- electrical forces (image charge, dipoles)

The image charge interaction is still important at distances of the order of several atomic diameters. At smaller distances, the classical treatment must be substituted by a quantum description of the interactions, introducing the exchange and correlation effects. Chemical interactions are the most significant contributions to the work function. The electrical components depend both on the chemical nature of the solid and on morphology of the surface. Indeed, in consequence of the local morphology, the charge redistributes itself yielding the formation of dipoles localized at the surface with which the expelling electron interacts.

It is now evident why the work function doesn't reach its theoretical value immediately outside the surface. This is a very important issue, because the

CHAPTER 2. Tunneling in electrochemistry

behaviour of the tunneling junctions depends on the extension of the gap which is usually in the range where these short distance effects dominate.

The simple overlap at different distances of two surface external potential profiles like that shown in fig. 2.3, generates tunnel barriers of different height (fig.2.4).

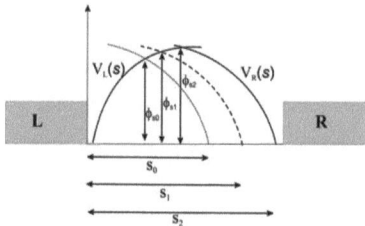

Figure 2.4: *Different barrier heights ϕ_i in a tunneling junction of different widths s_i.*

It should be stressed that in a tunneling experiment the barrier doesn't coincide with the work function as it is stated in some books. There is another fundamental difference between the two quantities: the work function is a thermodynamic parameter of the surface as an all, independent on the path followed to reach a position where the electron doesn't interact with the surface.

The tunneling barrier depends on the position and on the different paths that the electron can follows to reach the opposite side of the gap and this is what makes a tunneling barrier a local, extra-thermodynamic property which is far to be regular and symmetric.

The latter issue has important consequences on the interpretation of experimental data which, for the case of the present work are complicated by the presence of an electrolyte into the tunneling gap .

The direct measure of the *shapes* of the tunneling barriers at the electrified interface are presented in chapter 6, where the influence of both electrode potential and electrolyte is discussed.

2.2 Electron transfer at the electrified interface

The net current flowing through the interface at a given electrode potential is the sum of the anodic and cathodic process and it is expressed by the

2.2 Electron transfer at the electrified interface

Butler-Volmer equation:

$$i = i_o(e^{(1-\beta)F\eta/RT} - e^{-\beta F\eta/RT}) \qquad (2.11)$$

where i_o is called *exchange current* and represents the common value that both the anodic and the cathodic components assume at the equilibrium for $\eta = 0$ and $i = 0$ and β is the so called *symmetry factor*, which refers to the passage of the reacting system through an activated state.

Anyway, Eq. 2.11 doesn't say where the electron leaving the conduction band in the metal finally reaches the oxidized state, or where the occupied states which are left by the electron during an anodic process are localized.

It has been shown in sec. 1.3 that the interface electrode/electrolyte has a structure due to the distribution of the ions and water in a region extending for several molecular diameters apart. The closest approach for a solvated, not specifically adsorbed ion is at the Outer Helmholtz Plane (OHP). It's just the ions populating the OHP which are the most involved in exchanging electrons with the electrode.

As a consequence, the potential drop which actually determines the transfer rate is just that between the metal surface and the OHP, and represents only a fraction of the total potential drop through the interface. That gives rise to a correction term for the exchange current named after the electrochemist who was the first to be aware of the importance of the structure of the double layer in the electron transfer reactions, the *Frumkin coefficient*.

An early quantum-mechanical description of the electron tranfer process at the electrified interface in terms of tunneling process was proposed by Gurney [27].

In the quantum mechanical treatment, the rate constant W_{AD} for a electron transfer between a weakly interacting donor (D) and an acceptor (A) is broadly represented by the equation:

$$W_{AD} = |T_{AD}|^2 \left(\frac{\pi}{\hbar^2 \lambda KT}\right) \cdot exp\left[-\frac{(\lambda + \Delta G^0)^2}{4\lambda KT}\right] \qquad (2.12)$$

where λ is the reorganization energy of the solvent and $|T_{AD}|$ is the pre-exponential electron exchange factor which allows for a theoretical evaluation of the semi-empirical constants for the rates. This factor is represented by off-diagonal hopping integrals of the quantum-mechanical coupling matrix between A and D and contains the tunneling probability P_T which assumes different values for different tunneling mechanisms usually distinguished between localized transition called *adiabatic* (e.g. homogeneous electron transfer) and delocalized process (e.g. electrochemical charge transfer, STM) called *nonadiabatic* (ref. [28] and references therein).

CHAPTER 2.Tunneling in electrochemistry

The emerging field of the nano science and always new biological systems stimulate a new effort into the fascinating field of the electron transfer and its fully quantum-mechanical description. A topic which unfortunately must remain out of this work!

But how do the parameters which characterize the tunneling effect discussed in sec. 2.1 enter the description of the electrochemical processes?

From the discussion above, the electronation current density is given by:

- the number $N(E_F)$ of electron with Fermi Energy E_F that collide per second from inside the metal with a unit area of the metal/solution interface;

- the probability P_T that the electrons tunnel to the ions;

- the probability P_A that the system is in an activate state;

- the number N_{OHP} of ions populating the OHP;

i.e.
$$i_c = e \cdot N(E_F) P_T P_A N_{OHP} \qquad (2.13)$$

It can be shown that a displacement of the interface from the equilibrium by applying an overvoltage η has a direct effect on the activation energy of the electrochemical reaction and then on the probability P_A.

By assuming a tunneling barrier between the metal surface and the OHP having an arbitrary shape $\phi = \phi(s)$, the probability that the electron crosses the classically forbidden region is (see Eq. 2.10):

$$P_T = exp\left[-\frac{2}{\hbar}\int_0^{OHP}\phi(s)ds\right]^{1/2} \qquad (2.14)$$

In Eq. 2.14 the overvoltage doesn't appear. However, the overpotential η determines the properties of the liquid side of the interface and then indirectly influences the shape of the barrier $\phi(s)$ and all the quantities correlated. An experimental proof is given in chapter 6.

2.3 The Scanning Tunneling Microscopy (STM)

The STM is a device invented by G. Binnig and H. Rohrer for the study of the structure of surfaces and their electrical properties down to atomic and sometimes subatomic scale [29].

2.3 The Scanning Tunneling Microscopy (STM)

The exceptionally high resolution of the STM measurements opened a new era of the surface and material science, a revolution culminated in the contemporary nano-technology and nano-science.

In an STM a very sharp metal tip is approached to a surface till a tunneling current is established. The signal is hence processed by a computer and transformed into a topographic image.

The success of this technique stimulated the development of a series of Scanning Probe Microscopies, the most important being maybe the AFM (Atomic Force Microscopy) and somehow the SECM (Scanning Electrochemical Microscopy).

It appeared immediately clear after its invention, that the possibility to control the movements of a tip at the nano-scale, makes the STM not only a powerful method to image the surface but also to locally modify it. Very often, in the STM setup the tip serves not only as scanning probe but also as a nano-tool. This interesting opportunity given by the STM is one of the main topics of the present work and its application in electrochemistry is presented in chapter 5.

In recent years, increasing interest is arising around the characterization of the local electrical properties of surfaces, a goal mainly achieved with the *tunneling spectroscopy* (TS). An application of TS to the electrified interface is reported in chapter 6.

In spite of its success in the study of surfaces and of the deep knowledge of the theory behind, the STM must be still considered a developing technique [30]. In order to better understand what kind of informations can be extracted from a STM experiment, in the following some very fundamental theoretical aspects are discussed.

Based on the extreme sensitivity of the tunneling current to the tip/sample separation, in a STM the electrical signal is processed in order to give a topographic image of the surface. It is hence clear that an STM image is quite different than a photographic one: the STM is not an optical microscopy. That's why the interpretation of the STM images in terms of topography is not always that straightforward!

The technical description of the hardware setup of an STM can be found in [29, 30]. Let us here get closer to the theory behind.

A first order perturbation transfer Hamiltonian approach leads to the following expression for the current detected at the tip under the limit of small voltage and temperature [24, 31, 32]:

$$I = \frac{2\pi e^2}{\hbar} U \sum_{\mu,\nu} |M_{\mu\nu}|^2 \delta(E_\nu - E_F)\delta(E_\mu - E_F) \quad (2.15)$$

where U is the tunneling voltage between tip and sample, $M_{\mu\nu}$ are the ele-

CHAPTER 2. Tunneling in electrochemistry

ments of the tunneling matrix, E_μ, E_ν are the energies of the initial and final state and E_F is the Fermi level. The main problem is how to calculate the $M_{\mu\nu}$. The strategy proposed by Bardeen [24] already discussed for the general solution of the tunneling problem, can be used only if a wave-function for the tip is available.

Unfortunately, little is known about the structure of the tunneling probe tip which is prepared in a relatively uncontrolled and not reproducible way.

Tersoff and Hamann proposed to assume the tip to be locally spherical with a radius of curvature R where it approaches nearest the surface at a distance s from the curvature center r_0 as sketched in fig. 2.5.

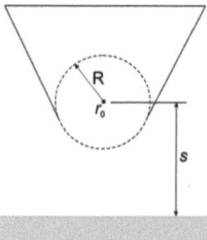

Figure 2.5: *Schematic representation of the geometry of the tunneling gap in the Tersoff and Hamann model.*

In agreement with the geometry of the spherical potential, a spherical s-type wave function is associated to the tip. With this assumption the current of Eq. 2.15 can be written as:

$$I \propto U \cdot n_t(E_F) \cdot exp(2\chi R) \cdot \sum_\nu |\Psi_\nu(r_0)|^2 \delta(E_\nu - E_F) \qquad (2.16)$$

where $n_t(E_F)$ is the density of states at the Fermi level of the tip, Ψ_ν are the wave-functions of the surface, χ is the local barrier height. In Eq. 2.16 the quantity:

$$\rho(r_0, E_F) = \sum_\nu |\Psi_\nu(r_0)|^2 \delta(E_\nu - E_F) = LDOS \qquad (2.17)$$

is nothing but the local density of states (LDOS) at the Fermi level of the substrate.

Hence, in the frame of the transition Hamiltonian theory, an STM image recorded in constant current mode represents a profile of constant density of states at the Fermi level of the substrate.

As I always said to students during the practical exercises of physical chemistry, what can be learned from an STM image is the electronic structure

2.3 The Scanning Tunneling Microscopy (STM)

of the surface, and in some case it can be very different of its topography, especially at atomic scale. And that is exactly what makes the difference between tunneling microscopy and photography!

Nevertheless, the approximation for the wave-function of the tip used here is too simple, and in the reality the expression of the tunneling current is not that simple as in Eq. 2.15.

Anyway, the method yields the following, extremely important expression for the tunneling current in an STM:

$$I_{STM} \propto \rho_s(s, eU) \cdot e^{-2\chi s} \quad (2.18)$$

where

$$\chi = \frac{(2m\phi(s))^{1/2}}{\hbar} \quad (2.19)$$

From Eq. 2.19, an *Effective Barrier Height*(EBH) can be defined as:

$$\phi(s) = \frac{\hbar^2}{8m}\left(\frac{dlnI}{ds}\right)^2 \quad (2.20)$$

An alternative way for presenting the STM results is to report profiles of constant barrier height $\phi(s)$ [30].

An other important implication of Eq. 2.18 is that the quantity dI_{STM}/dU should give a direct measure of the density of states of the sample.

In principle, due to the high spacial resolution of the STM, by measuring the current for fixed tip/surface distance at different bias, a *picture* of the LDOS at different places of the surface could be achieved.

Unfortunately, as it will be extensively discussed in chapter 6, due to the high degree of approximation of the Tersoff and Hamann model, the LDOS can't be identified with the quantity dI_{STM}/dU although the experimental determination of this parameter is important for the study of the electronic structure of the surface.

The Tersoff and Hamann theory has been a powerful tool for the interpretation of STM results of simple surfaces such as Au(110) or Si(111), but, as reported by Tersoff self [34] and successively followed by other authors [35–37], the transition Hamiltonian approach is unable to reproduce experimental results for close packed surfaces as Au(111), Al(111) and for HOPG.

Indeed, in this theory developed for low voltage and temperature [31,32], an STM image is only due to the electronic features of the surface neglecting the electronic structure of the tip, and moreover the interaction between tip and substrate.

By experience, I can say that the choice of the metal-tip, which is usually made by transition metals, influences the quality of the STM images and one

CHAPTER 2.Tunneling in electrochemistry

should not wonder that also the geometry of the gap plays a role, although the latter is unknown.

In these experiments the nature of the tip plays a determining role in the correct reproduction of the surface corrugation: the s-wave approximation is inappropriate.

In order to give a theoretical basis to the experimental results for close packed surfaces and to supply scientists with a tool where the algorithms of the density functional theory and of the quantum chemical methods can be implemented, a completely different approach was proposed by Marcus and co-workers [38], who treated the problem of the tunneling current in an STM with a two Hamiltonians formalism. In this approach, the tip is modeled as a semi-infinite chain of atoms treated in tight-binding approximation, where atom-centered orbitals can be used for the valence band of the tip.

This is the main difference with the model of Tersoff and Hamann: when the atom of the semi-infinite chain are transition metals (and that is in practice always the case), five not interacting d-orbitals are used.

Whit this assumption, an expression is obtained where the current depends explicitly on the coupling of the tip-states with those of the sample.

The model gives the correct calculations of the STM images for graphite and Au(111), and also explains the difference in the experimental corrugation when different tip materials are used [39].

More important, the two Hamiltonians theory is suitable to be refined in order to give a correct description of the image of adsorbates.

The reacting states of an adsorbate are usually the HOMO and/or LUMO molecular orbitals which are usually expressed as a linear combination of single centered atomic orbitals. In agreement with the Marcus treatment, the different contributions, namely the different coefficients of the linear combination determine the local brightness at the different sites of the molecule. Negative contributions to the MO usually yield dark contrast in the STM image and vice versa.

Furthermore, the model predict that the tunneling current is also mediated by atomic or molecular states which may take significant values in the space between tip and substrate, or be strongly off-resonance but close to the target atom [41]. Since the relative energy of the states involved in the tunneling process can be shifted by applying a bias, the theory gives also an account for the bias-dependent STM images of adsorbates[40]. This kind of contributions to the total tunneling current which describes the corrugation of the STM image in terms of tip effect were not possible to describe within the theory of Tersoff and Hamann.

At the electrified interface, the solvent polarization will significantly modify the local charge distribution of the adsorbate, and the orbital involved in

2.3 The Scanning Tunneling Microscopy (STM)

the tunneling resonance will be not the same than those involved in vacuum.

That accounts for the differences between *in situ* and *ex situ* experimental results [42].

CHAPTER 2. Tunneling in electrochemistry

Chapter 3

Experimental

3.1 STM

The results reported in this work are all obtained with a commercial Nanoscope E III STM from Digital Instruments, Santa Barbara California.

In order to perform *in-situ* experiments, the disk electrodes occupied entirely one side of small electrochemical cells which avoid the leakage of electrolyte by mean of a sealing ring.

The cells are made of Kel-F (polytrifluoroethene), a material suitable to be cleaned in very aggressive acidic media.

Indeed, all the glassware and cells used, were cleaned by immersing them in a concentrate solution H_2SO_4/H_2O_2 3:1. This very oxidizing solution allows for the removing of all impurities. The residual acid is washed out with boiling pure water (Milli-Q, 18.2 $M\Omega cm$).

As reference and counter electrode a Pt wire was used. Although this quasi-reference electrode is characterized by small changes of the potential during the measurement time($\approx +500mV$ vs. SCE), it allows for a high degree of cleanness. The Pt wires can be cleaned with the acidic solution or by annealing with an hydrogen flame. Both procedures were used.

The tips were obtained with the standard etching methods of 80:20 Pt/Ir wires in a 4M NaCN solution. The wire served as anode in an electrochemical system closed with a Pt ring (cathode) which sustains a lamella of the etching solution. The system was given a potential of 4 V and the film-electrolyte was changed every 2 minutes by dipping the ring in the solution.

The residual salt was removed from the etched wire by washing with pure water.

In order to minimize the faradaic current at the tip, which would make the *in-situ* measurements impossible, the tips were insulated by deposition of a polymeric electrophoretic paint (Electrocoating ZQ8-43225, BASF, Lud-

CHAPTER 3.Experimental

wigshafen, Germany) by applying to the tip a potential of +80V for 3 minutes. The coat was dried in an oven at 200°C for 10-20 minutes. This treatment yields to a leakage current of $\approx 40pA$.

3.2 Electrodes

3.2.1 Au(111)

The electrodes were disk-Au single crystals with a diameter of $\approx 8mm$ (Matek, Jülich, Germany). Before each measurement the electrodes were flame annealed for 5 minutes at red heat and cooled down in a N_2 atmosphere. This preparation method should remove contaminants eventually present at the surface and make it more regular.

Indeed, in order to completely remove metal deposits from the active surface, the electrode were electropolished by oxidation in H_2SO_4 0.1M at +10V for 5-10 seconds. This leads to the formation of an oxide on the gold surface which is removed by immersing the electrode in HCl 1M for at least 3 minutes. Finally the electrode was washed with pure water.

This procedure, while allows for the complete removing of the deposits, causes a roughening of the surface because the electropolishing is achieved by removing gold layers. Several cycles of flame annealing restore the surface quality.

For the SECM experiments described in chapter 5, bead single crystals obtained with the method developed by Clavilier were used [121, 122]. With this method, small single crystals are obtained by gradually melting of a gold wire. By cooling the gold drop very slowly, single crystals of very high quality can be produced. The contaminants are removed by treating the drop with aqua regia (HCl/HNO_3 conc. 3:1).

The Au(111) facets of the single crystals obtained with the Clavilier methods are clearly visible with an optical microscope. A photograph is shown in fig. 5.27. When observed with the STM, these facetes show very large terraces extending for micrometers.

Once the Au(111) facet was found on the bead, the crystal was immobilized with the desired orientation by fixing it on a gold foil. Before the use, the bead-crystals were electropolished and flame annealed.

3.2.2 Ir(210)

The Ir(210) crystals were prepared by K. A. Soliman, Institute of Electrochemistry, University of Ulm.

The Ir(210) crystal of 4mm diameter (Mateck, Jülich, Germany) was annealed by inductive heating at 1000°C for 30 seconds in an hydrogen atmosphere for the preparation of the faceted surface, and in a mixture of hydrogen and nitrogen in order to obtain the planar surface.

The electrode was then slowly cooled down to room temperature, and the quality was checked by measuring a cyclic voltammogram immediately after the annealing.

During the transfer into the STM-cell, the surface was protected with a drop of pure water.

3.3 Nanostructuring

The electrochemical nanostructuring was achieved with a fully automated system developed at the Institute of Electrochemistry, University of Ulm [111].

The controller of the STM was modified with an electronic card which is able to send pulse-voltages to the three directions of the piezo elements of the STM head while the microscope images the surface.

The card is controlled with an external software which allows to change the following parameters:

- Pulse voltage: this parameter control the pulse sent by the system to the z-direction of the piezo element in order to approach the tip to the surface

- Pulse length: is the duration of the voltage pulse. Usually 4-5 milliseconds

- Waiting time: is the time the tip is kept away from the surface in order to get loaded with sufficient material. It changes for different systems and depends on the metal deposition rate onto the tip

- Distance between the clusters of a single line

- Distance between lines

All the measurements reported in chapter 5 are obtained by fixing these parameters.

CHAPTER 3.Experimental

3.4 Tunneling Spectroscopy

For the tunneling spectroscopy measurements, the oxygen can influence the results and must be removed.

The STM was placed in a sealed box, where an argon atmosphere was maintained. Any presence of oxygen was detected by cyclic voltammetry (see fig. 6.15).

Several tip and sample cyclic voltammograms were executed before each experiment in order to electrochemically stabilize the system.

The tunnel current-distance $(I_T - s)$ curves were recorded by first setting 800nA the current set point in order to fix the distance. This is the maximum value that could be handled by our modified tip preamplifier.

Then the feed-back loop of the STM was switched off and $I_T - s$ traces were recorded while the tip was retracted from the surface and approached again at a rate between 2.3 nm/sec and 6 nm/sec, a velocity that minimize thermal drift effects from the STM head.

Then the feed-back was briefly reactivated to check and eventually correct the position on the surface, before two new curves (forward and backward) were recorded.

The measurements were repeated at different places of the surface. The measurements were accepted only when the tip gave a good atomic resolution before and after the I_T-s measurements and when the forward and backward currents, within given limit, were identical.

The z-calibration of the scanner was checked before and after each session. The curves reported in chapter 6 are average of 500 curves recorded with different tips and on different days.

In order to calculate the Effective Barrier Height directly from the measured currents, uniformly spaced $I_T - s$ data points were produced with a cubic spleen algorithm without changing their total number [43]. This procedure allows also for a filtering of any noise arising from the electronics, from faradaic currents at the tip and from the stepped movement of the piezo. An example of the effect of this procedure on the experimental data is shown in fig. 3.1.

For the voltage spectroscopy, after the set-point current was set at the desired value, the feed-back loop was switched off and the sample potential was linearly changed within the desired range at a rate of 0.1 V/sec.

The tip was held at a fixed potential while the current was recorded during a complete cycle of sample potential.

The the feed-back was shortly reactivated in order to check and eventually correct the current set point.

The tip had to fulfil the same requirements of the DTS measurements.

3.5 Chemicals

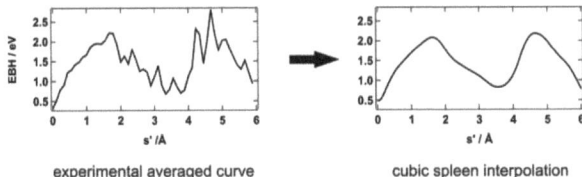

experimental averaged curve   cubic spleen interpolation

Figure 3.1: *Effect of the cubic spleen interpolation algorithm on the experimental data.*

Uniformly spaced $I_T - V$ curves were generated with a cubic spleen algorithm directly from averaged (20-50 curves) experimental curves recorded with different tips and on different days.

3.5 Chemicals

Chemical	Producer	use
H_2SO_4 96%, suprapur	Merck	electrolyte
HCl 30%, suprapur	Merck	electrolyte
$CuSO_4 \cdot 5H_2O$, p.a. >99%	Merck	electrolyte
$PdSO_4$, p.a. >99%	Alfa Aesar	electrolyte
H_2SO_4 96%, p.a. 95%-97%	Merck	cleaning solution
H_2O_2, med. reinst. 35%	Merck	cleaning solution
$Pt_{80}Ir_{20}$ 0.25 diam.	Degussa AG	STM tip
ZQ8-43225 Electrophoretic paint	BASF	Tips insulation
Kel-F	Heute & comp.	STM cells

CHAPTER 3. Experimental

Chapter 4

Nano-faceting of surfaces: the case of Ir(210)

4.1 Introduction

The instability of high-index single crystal surfaces has been a field of intense research for decades. The surfaces of body-centered cubic (111) and face-centered (210) crystals undergo morphology changes in the presence of strongly interacting adsorbates described in the literature in terms of surface *reconstruction* and *faceting* [44–46].

In the case of faceting, an initially flat surface is replaced by a another one with a lower total surface energy and a larger extension due to the formation of facets with different orientations.

The driving force for such transformation is a strong surface energy anisotropy which is responsible for the evolution of the system (the surface) toward a thermodynamically more stable state in a way which is in close relation with the transitions between thermodynamic phases like, *e.g.* liquid and gas [47].

In agreement with the treatment of Herring [48], the thermodynamical condition for the transition from an initial, flat surface to a faceted one is:

$$\sum_k \frac{S_k}{cos(\theta_k)} \gamma_{k,fac} - \gamma_{ini} < 0 \qquad (4.1)$$

where γ_{ini} and $\gamma_{k,fac}$ are the formation energy per unit area respectively for the initial surface and for the k-th facet forming an angle θ_k with the flat surface. S_k is the normalized surface, a parameter which accounts for the extension of the k-th facet.

Anyway, the relation of Eq. 4.1 is a necessary condition, but is not sufficient for the faceting. Indeed, open surfaces with an high density of

CHAPTER 4.Nano-faceting of surfaces: the case of Ir(210)

broken bonds don't facet spontaneously, also in the case of an high energy difference between the initial and the final state.

On the other hand, simple increase of temperature increases the energy isotropy, and at least could stabilize the unfaceted surface [49].

As already mentioned, the presence of an adsorbate is essential for the faceting process. Only by annealing the sample in the presence of an agent strongly interacting with a surface orientation different of the flat one, the kinetic barriers for the faceting can be overcome. It is not a surprise that the relaxation of a faceted surfaces to the initial unstructured one has been observed upon removing of the adsorbate [50].

Changes in the surface structure have been observed also during surface reactions involving strong interactions of either the reagents or the products with the surface [44].

This short and not exhaustive introduction points to the character of *softness* of such surfaces, and suggest the possibility to control or even tune the structure transformations down to nano-scale, and this is one of the most important aspects attracting the attention of the researchers since the catalytical behaviour of these systems seems to be caused by nanometer-scale effects [51–53].

For the same reasons, faceted surfaces are considered to be of major interest for electrochemical experiments where the structure of the surface is crucial for the behaviour of the entire interface.

In the following, an *in-situ* STM study of a faceted surface stemming from an Ir(210) is presented, where besides the morphological characterization, the effects of potential and electrolyte composition are investigated.

4.2 Morphology and faceting mechanism

Iridium is a metal of the platinum group with an fcc bulk structure. In order to study the catalytical and mainly the electrocatalytical behaviour of this metal, a detailed description of the surface is necessary.

It is already known from UHV experiments that the very open and rough Ir(210) surface orientation gets faceted upon annealing in presence of oxygen [46, 50, 54].

The results shown in the following refer to an Ir(210) crystal which was first characterized electrochemically after the faceting procedure before it was transferred into the *in-situ* STM.

As shown in fig. 4.1, the surface consists of pyramids with facets corresponding to different Miller indexes.

A three-dimensional image is shown in fig. 4.2.

4.2 Morphology and faceting mechanism

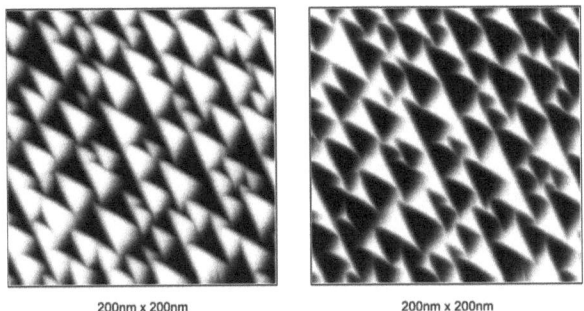

Figure 4.1: *STM image (left) and the relative inverted one (right) of a faceted Ir(210) surface. E=+440 mV/SCE, electrolyte: H_2SO_4 0.1M.* (In the inverted image, the white features corresponds to valleis in the real morphology of the surface).

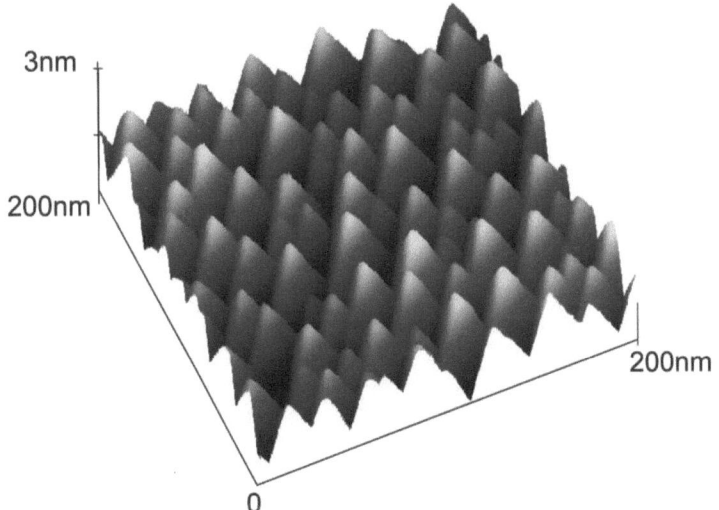

Figure 4.2: *3D STM image of a faceted Ir(210) in H_2SO_4 0.1M. E=+400mV*

CHAPTER 4.Nano-faceting of surfaces: the case of Ir(210)

In earlier studies, the pyramids have been already described to be made by two longer (311) facets and one smaller (110) [46, 50].

It can be seen that the pyramids are uniformly distributed at least within the imaged surface. The inverted (negative) image in fig. 4.1 (right) has been shown in order to better highlight the topography of the substrate.

A side view of the STM image in fig. 4.1 is shown in fig. 4.3. It can be seen that there are not planar zones alternating with the pyramids. The vertical distance between two consecutive extremes (apex/valley) never exceeds 3nm and the average width a the basis of the pyramids is 30nm.

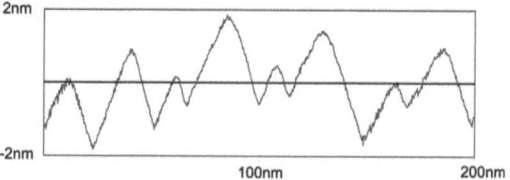

Figure 4.3: *Section analysis of fig. 4.1*

It must be said that the *size* of the pyramids changes slightly from day to day of measurement, pointing out that little differences in the preparation procedure (*e.g.*, different annealing times and/or temperatures) influence the the faceting process. In order to get more detailed information about the shape of the pyramids, which is alway reproduced, the first derivative of the STM images have been calculated and they are shown in fig. 4.4.

This kind of images has the advantage to better describe the irregularities and slope changes on the substrate, although the height is not anymore correctly reproduced. Indeed, fig. 4.4 reveals that the shape of the facets is square instead of triangular as it was thought [46, 54]. This is the consequence of the pyramidal geometry of the valley: a perfect triangular face would be possible if planar zones alternate with the pyramids.

Anyway, in the following the model proposed by Madey and co-workers [50] which describe the geometry of the pyramids as made by two (311) and one (110) facets is assumed to correctly describe the experimental results reported here.

A larger image is reported in fig. 4.5. The image was recorded in $HClO_4$, an electrolyte which is supposed to don't specifically adsorb, but under the present experimental condition no difference has been detected from the measurement in H_2SO_4 (fig. 4.1). Indeed, the electrode comes in contact with the solution at a potential where the voltammograms show only capacitive current.

4.2 Morphology and faceting mechanism

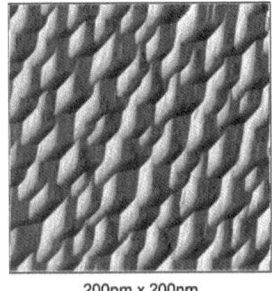
200nm x 200nm

50nm x 50nm

Figure 4.4: *First derivative images of the STM image shown in fig. 4.1*

700nm x 700nm

Figure 4.5: *STM image of a faceted Ir(210) surface in $HClO_4$ 0.1M. E=+400mV/SCE*

CHAPTER 4. Nano-faceting of surfaces: the case of Ir(210)

The STM measurements give also the opportunity to estimate the tilt angles (indicated in fig. 4.6) that the different facets form with the horizontal plane of the STM images. The (311) facet forms a tilt angle $\beta = 18\pm2°$, which reproduces the values obtained with different techniques already reported by Madey and co-workers [54]. For the other orientation, the angle results to be $\gamma = 9 \pm 2°$, a value which indicates the presence of a superstructure formed by the Ir atoms of the top most plane of a reconstructed (110) facet [50].

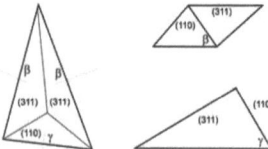

Figure 4.6: *Schematic representation of the tilt angles of the facets of the pyramids*

By carefully watching the disposition of the pyramids, some order can be seen. Indeed, the pyramids seem to share one edge along which they align. This common edge is very clear in the inverted image of fig. 4.1 which in the model proposed by Madey corresponds to the (311) direction.

In a different measurement reproduced in fig. 4.7 this situation is clearly visible and maybe is a consequence of the faceting mechanism.

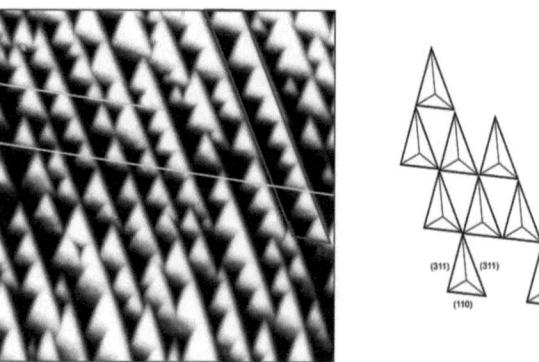

200nm x 200nm

Figure 4.7: *STM image of a faceted Ir(210) surface in $HClO_4$ 0.1M. E=+400mV/SCE. Right: model for the pyramids disposition.*

4.2 Morphology and faceting mechanism

A question that must be answered is whether the faceting procedure used yields a completely faceted surface or not. The STM measurements reveal that the faceted domains extend over a region usually not larger than $4\mu m$, before a different surface structure is imaged. The images of portions of the surface surrounding a completely faceted domain are shown in fig. 4.8. It can be seen that the pyramids decrease their size turning in ridges, ridges which relax to a flat surface.

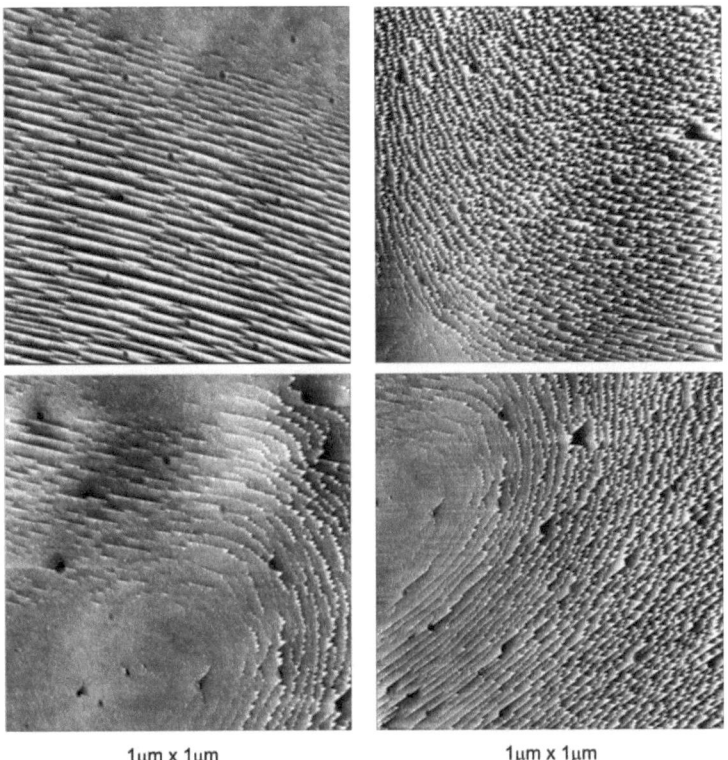

Figure 4.8: *STM images of 4 regions of a faceted Ir(210) surface surrounding a completely faceted domain. Electrolyte: $HClO_4$ 0.1M. E=+400mV/SCE*

Another example of what can be called a *transition region* from the faceted surface to the unstructured one, is shown in fig. 4.9, which actually

CHAPTER 4. Nano-faceting of surfaces: the case of Ir(210)

is the STM image of the surface surrounding the domain with the pyramids shown in fig. 4.5. The degradation of the pyramids to ridges is again evident. This seem to be a general property of the surface and during the STM measurements the pyramids were found just by following the ridges.

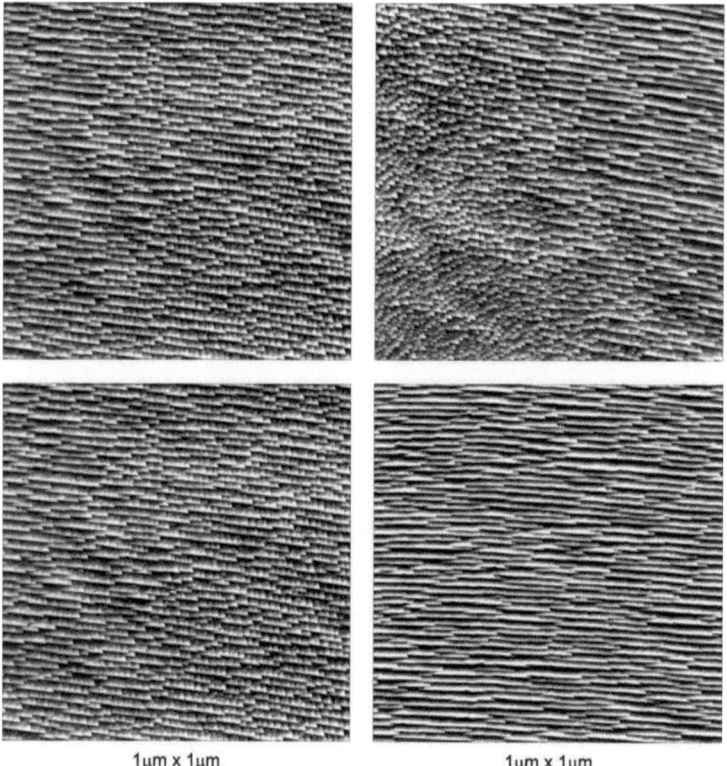

Figure 4.9: *STM images of 4 regions surrounding a completely faceted domain of a Ir(210) in $HClO_4$ 0.1M. E=+400mV*

Hence, the entire electrode surface appears to consist of a sequence of these large domains (faceted and transition) which have the same extension.

This is consistent with a growth of the pyramids which nucleate on different sites of the surface and maybe, depending on the preparation procedure,

4.2 Morphology and faceting mechanism

the different growing domains merge in order to completely facet the surface. In the present case the coexistence of flat, partially and totally faceted surface is evident.

If this results leads to the conclusion that the preparation protocol followed for the faceting of a flat Ir(210) surface is not efficient, on the other side could give the possibility to better understand the mechanism of the pyramids formation, just by considering the STM image of the transition region.

A detail of the STM image of the ridges already seen in fig. 4.8 is shown in fig. 4.10. The ridges have an average height of 2nm with the tilt angles (indicated at the bottom of the fig. 4.10) $\beta = 13 \pm 2°$ and $\alpha = 4 \pm 2°$. β

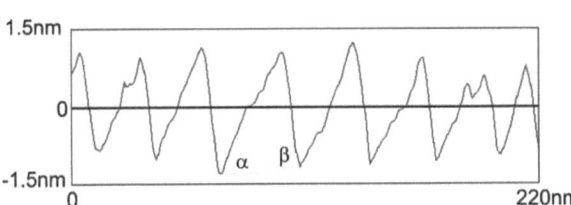

Figure 4.10: *STM image of a partially faceted Ir(210) surface in $HClO_4$ 0.1M. E=+400mV/SCE*

CHAPTER 4.Nano-faceting of surfaces: the case of Ir(210)

increases in direction of the faceted field. The height of the ridges and then α, decrease till zero as the surface becomes more flat.

The presence of the ridges in the transition region could give an explanation for the final alignment of the pyramids. A direct proof for this hypothesis is shown in fig. 4.11. In this image the ridges can be seen turning progres-

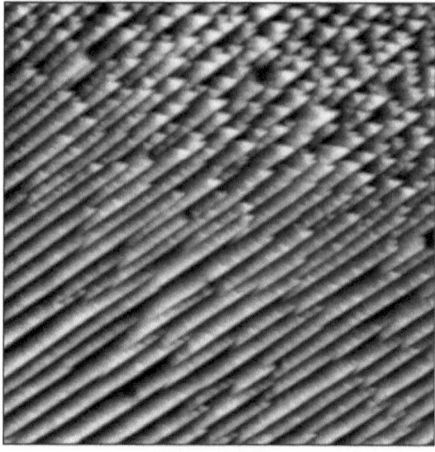

490nm x 490nm

Figure 4.11: *STM image of a partially faceted Ir(210) surface in $HClO_4$ 0.1M. E=+400mV/SCE*

sively into pyramids by formation of the different facets. It is interesting to note that the alignment of the pyramids follows the direction of the original ridge, and from the STM image it seems that the steeper edge of the ridges evolves toward the (311) facet of the final pyramid. If this is true, the adsorption of oxygen on this facet triggers the entire faceting process.

By summarizing,

- The faceting starts with the formation of the ridges by adsorption of a faceting agent;

- The ridges have a face which has an orientation evolving to the final (311);

- Th edge of the ridge is retained till a true (311) facet is formed. The faceting is completed with the consequent formation of the other facets

4.2 Morphology and faceting mechanism

from the collapse of the large surface of the ridge with a *domino-like* mechanism.

The model is sketched in fig. 4.12.

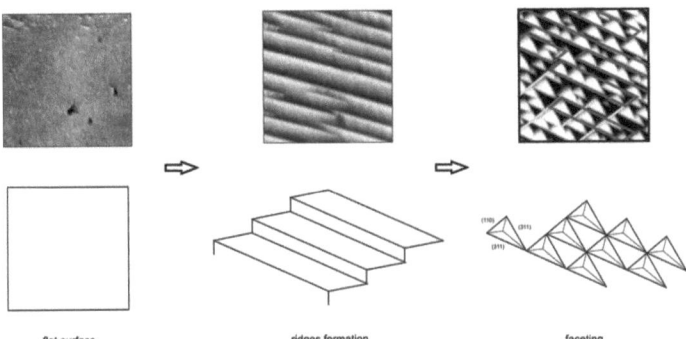

Figure 4.12: *STM image and schematic representation of the possible faceting mechanism of Ir(210)*

When compared with other methods which usually give an average description of the surface [54], the STM has the advantage to probe the local morphology.

Scanning large portions of the surface electrode revealed also the presence of defects. The presence of steps was detected, as shown in fig. 4.13.

What is exactly the origin of such surface morphologies is difficult to say from the STM image. The abrupt change in the structure which reveals clear borders between different regions can be associated to the presence of steps already on the planar surface. This is very evident in the bottom-right image.

Missing pyramids in a completely faceted domain is shown in fig. 4.14. The size and the position of such defects are randomly distributed and can be argued that holes were already present on the original surface. On the other hand, the annealing time and temperature could cause these features to appear.

It is well known that surfaces which undergo to faceting are very sensitive to the preparation conditions which strongly influence the process: in fig. 4.15 is reported the STM image of a faceted Ir(210) obtained after annealing upon not well established conditions where the contamination of the annealing gases was suspected.

CHAPTER 4. Nano-faceting of surfaces: the case of Ir(210)

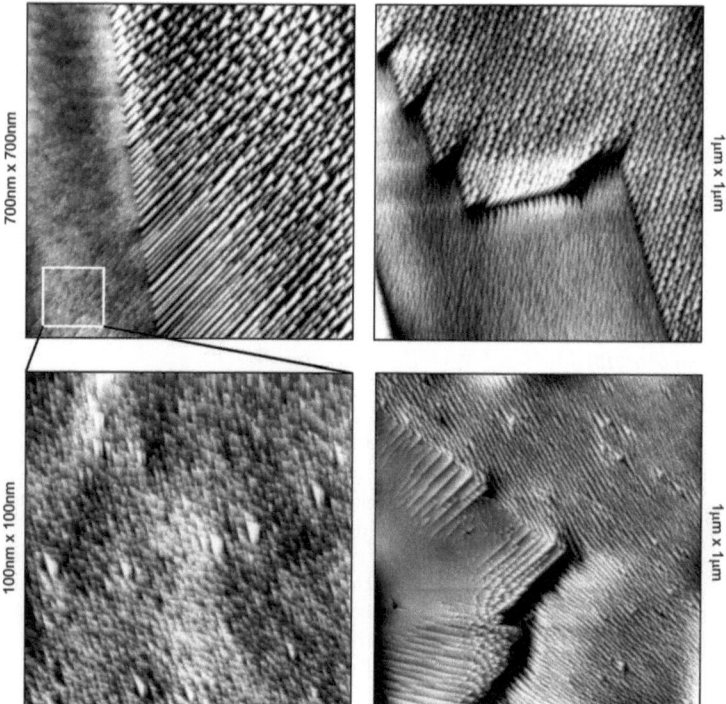

Figure 4.13: *STM images of different portion of a partially faceted Ir(210) surface in $HClO_4$ 0.1M. E=+400mV/SCE*

4.2 Morphology and faceting mechanism

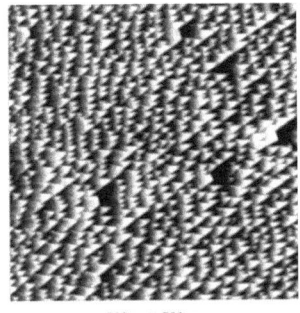

Figure 4.14: *STM image of a faceted Ir(210) surface in $HClO_4$ 0.1M. E=+400mV/SCE. Missing pyramids defects can be recognized.*

Figure 4.15: *STM image of a faceted Ir(210) surface in $HClO_4$ 0.1M. E=+400mV/SCE. This structure was find after annealing under not well defined conditions where the presence of contaminants in the annealing gas was suspected.*

CHAPTER 4.Nano-faceting of surfaces: the case of Ir(210)

4.3 Electrochemical behaviour of the faceted Ir(210)

(*The voltammograms shown in this section have been measured by Kahled A. Soliman, Institute of electrochemistry, University of Ulm*)

The study of the morphology of an electrode is necessary for an interpretation of its electrochemical behaviour. In the previous section the STM has been employed in order to check the preparation of faceted Ir single crystals and for a determination of the very complex structure of the electrodic surface.

A cyclic voltagram for a planar, not faceted Ir(210) surface is shown in fig. 4.16.

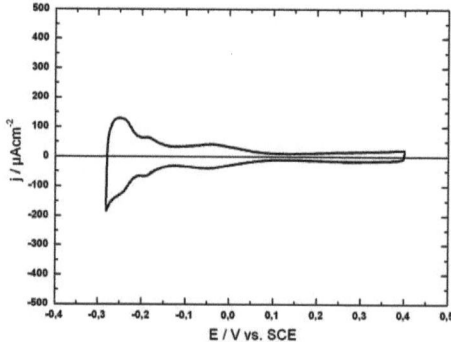

Figure 4.16: *Cyclic voltammogram for a planar Ir(210) in H_2SO_4 0.1M. Scan rate: 50 mV/s. (Courtesy of K. A. Soliman)*

Even without entering into the details, the dramatic consequences of the nano-faceting of the flat surface on the behaviour of the electrode can be seen by comparison of the voltammogram reported in fig. 4.16 with those shown in fig. 4.17 and in fig. 4.18 which have been performed respectively in 0.1M H_2SO_4 and 0.1M $HClO_4$.

The extremely complex structure of the surface makes not easy to explain all the features appearing in the voltammograms, which hence represent an overimposition of signals coming from the different crystal orientations of the

4.3 Electrochemical behaviour of the faceted Ir(210)

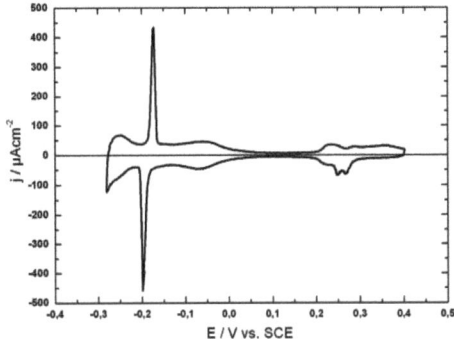

Figure 4.17: *Cyclic voltammogram for a faceted Ir(210) in H_2SO_4 0.1M. Scan rate: 50 mV/s. (Courtesy of K. A. Soliman)*

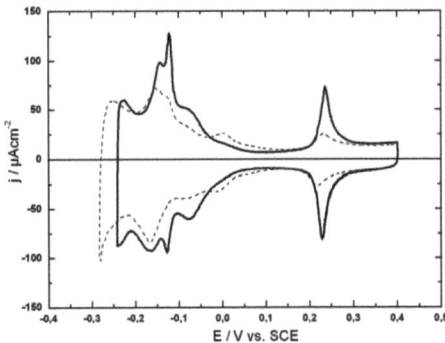

Figure 4.18: *Cyclic voltammogram for a faceted Ir(210) in $HClO_4$ 0.1M. Dashed line: same electrode after 2 hours. Scan rate: 50 mV/s. (Courtesy of K. A. Soliman)*

CHAPTER 4.Nano-faceting of surfaces: the case of Ir(210)

nano-facets together with those originating at the unstructured and partially faceted regions which have been observed with the STM.

But this aspect is also what makes this kind of surfaces very interesting since the *special* morphology of the faceted surface allows for localized process at the nanoscale.

An example is given by the behaviour of the faceted Ir(210) in sulfuric acid. Sulfate is an anion which specifically adsorb on several surface of noble metal and the sharp peaks appearing in the voltammogram of fig. 4.17 at -0.2 V/SCE can be assigned to a process of sulfate adsorption/hydrogen desorption.

In fig. 4.19 an STM image recorded at a potential positive of that of the two peaks, reveals the presence of an ordered adlayer on only one facet of the pyramids. This adlayer is better visible in the first-derivative-STM image on the right. It is evident that sulfate builds an ordered adlayer selectively on the (110) facet.

In order to make sure that what has been imaged is really a sulfate adlayer and not the Ir(110) superstructure of the reconstructed facet, the tilt angle has been estimated with the STM: quite surprisingly the angle results of $20 \pm 2°$. Although the presence self of the adlayer can influence the apparent angle, the measured value rules definitively out that what is shown in fig. 4.19 is the superstructure.

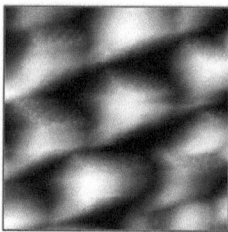

50nm x 50nm   50nm x 50nm

Figure 4.19: *STM image (left) and its first derivative (right) of a faceted Ir(210) surface. An ad layer can be oserved on one facet. E=+400mV/SCE. Electrolyte: H_2SO_4 0.1M*

No evidence of a formation of ordered adlayer was found in $HClO_4$.

The double-peaked feature appearing in the voltammogram of fig. 4.18 at ≈-150 mV/SCE are most likely due to hydrogen adsorption on the (110) facet. Very similar peaks have been alredy reported on Ir(110) [55, 56] and Pt(110) [57] bulk crystals.

4.3 Electrochemical behaviour of the faceted Ir(210)

The (311) orientation gives also a sharp peak around -100 mV/SCE, and as already said, the behaviour of the electrode in that range of potentials can be due to the concomitant processes occurring on the different facets.

The peaks appearing up to +200 mV/SCE are most likely due to adsorption/desorption processes involving OH^- [58].

This could also explain the variation of the voltammogram of fig. 4.18 (dotted line) with the time, which could be due to the competition for the adsorption of OH^- with Cl^- which is formed by reduction of $HClO_4$ [59], since no changes in the morphology have been observed with the STM.

The deposition of Cu on the faceted surface of Ir(210) has been also studied.

In fig. 4.20 a jump in the electrode potential causes a deposition of Cu-clusters on not well identified places of the surface.

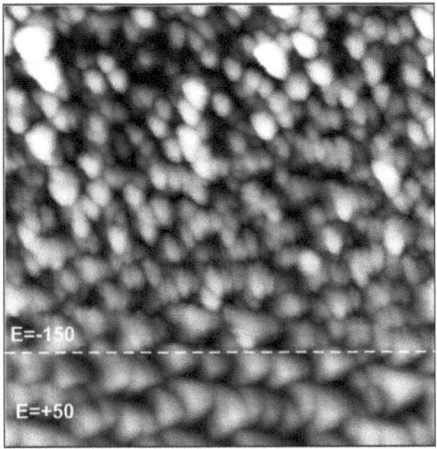

Figure 4.20: *STM image of a faceted Ir(210) surface at two different potentials. Electrolyte: H_2SO_4 0.1M + $CuSO_4$ 1mM*

By carefully controlling the potential variation, the Cu deposition seems to start on the small (110) facets of the pyramids as shown in fig. 4.21, where again the first-derivative STM images has been calculated in order to highlight the features of the surface.

Although it is not yet fully understood, the results presented in this section demonstrate that the behaviour of a faceted Ir(210) is due to a series

CHAPTER 4.Nano-faceting of surfaces: the case of Ir(210)

of processes occurring at the nano-scale level.

Moreover, the question of a potential induced transformation of the surface morphology could be arisen. In the measurements presented above, which must be considered as a preliminary investigation of the electrochemical behaviour of faceted Ir(210) electrodes, no influence of potential and electrolyte on the surface structure has been observed.

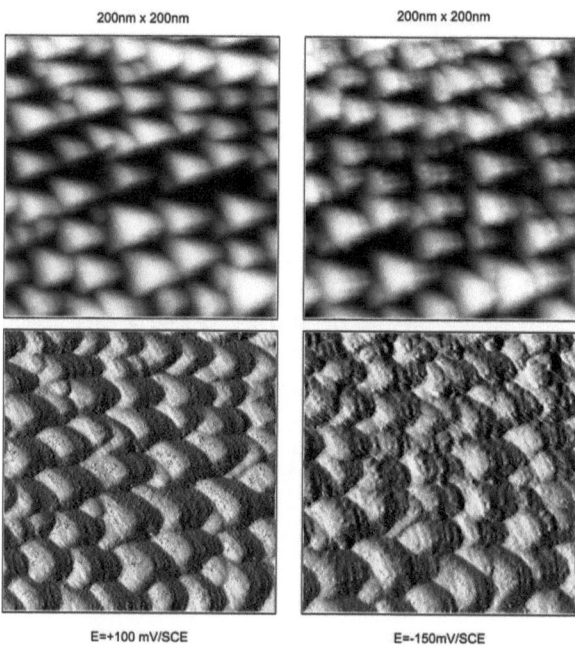

Figure 4.21: *STM images (left) and their first derivative (right) of a faceted Ir(210) surface a two different potentials. Cu deposition seems to start at the (110) facets. Electrolyte:* H_2SO_4 *0.1M+* $CuSO_4$ *1mM*

Chapter 5

Electrochemical nanostructuring with the STM

5.1 Introduction

The STM is one of the methods intensively used in electrochemistry to obtain detailed informations about the topography of the electrode surface. It's not a surprise that the STM has been a breakthrough in the building of an electrochemical surface science, which is one the leading topics of contemporary electrochemistry [60, 61].

After its invention, it turned soon clear that the STM offers the possibility to get beyond the *simple* aim of imaging a surface because its special setup allows to handle the tip as a tool which can be controlled at the nano scale.

Although the first successful experiments were conducted in UHV [62–68], it is in electrochemistry that the use of STM for nano-modification of the surfaces is widely used [71, 72] because its versatile setup allows to work at room temperature with ionic solutions which are actually an infinite reservoir of ions. And these are just some of the advantages over the UHV alternative.

Many of the strategies of contemporary *in-situ* nanostructuring are sketched in fig. 5.1. These methods are shortly reviewed in the following where their intimate electrochemical nature is highlighted.

Based on the well known behaviour of surface defects as nucleation centers for metal deposition, Penner and co-workers used an STM for the local generation of nano-sized defects just by crashing the tip against the surface at a wished position (Fig. 5.1a) [73–75]. The process can be repeated at will and the surface can be decorated with pre-designed patterns of metal nano-clusters.

If at the beginning this strategy has been used for the nano-modification of exclusively metal surfaces, during the last years Homma applied the same

CHAPTER 5. Electrochemical nanostructuring with the STM

method for the decoration of Si substrates just by exposing the samples patterned with nano-defects to solutions containing metal ions. Example of Cu, Ag and Au clusters on silicon were reported [76–78].

By acting on the piezo-crystal of the STM-head, the tip can be brought very close to the surface and make it to penetrate a tarnishing film causing an high overpotential for the metal deposition (Fig. 5.1b). By further moving laterally the tip, the film can be locally removed and if the sample is held at a potential slightly negative of the Nernst equilibrium value, metal deposition occurs only on the freed portion. In this way Cu has been deposited on CuO layers [79] and on Au(111) covered with a SAM [80].

It must be said that bringing the tip very close to the electrode causes the merging of the double-layers of both sample and tip. In this case tip and sample potentials are not anymore independent, which is the basic assumption for the STM, so that the electrical contact with the reference electrode is lost and the local electrochemical condition are not anymore well defined. This configuration is defined as *double layer cross talk*, a particular electrochemical configuration which has been used either for local dissolution (fig. 5.1d) and for local deposition (fig. 5.1c).

In the first case, Kolb and co-workers have reported a tip induced local dissolution of a Cu surface while the electrode is held at a potential where no

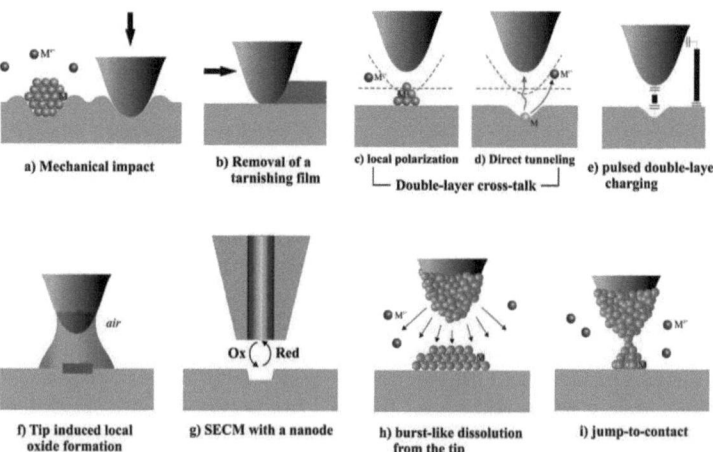

Figure 5.1: *Schematic representation of different strategies for the electrochemical nanostructuring with the STM.*

5.1 Introduction

dissolution is expected [81]. The phenomenon has been explained by assuming the direct tunneling from the Cu/Cu^{++} redox couple to the STM tip. This tip-induced local dissolution has been then used for local modification of semi-conductor surfaces, where the same mechanism seems to be active [82, 83].

An example for local deposition induced by local polarization of the electrode with the tip was given by Widmer and co-workers (Fig. 5.1c) [84]. In this case, after the tip has been brought in close proximity of the surface, a pulse of anodic overpotential is applied. The cluster deposition occurs in a local cathodic process at the surface presumably via a local tip-induced negative image charge.

The time-charging properties of the electrochemical double-layer have been used by Schuster and co-workers for the controlled etching of surfaces by applying a nanosecond voltage pulse between tip and substrate (Fig. 5.1e) [85–87]. Due to the conductivity of the solution, during such a short time only the double layers of the nearest region between tip and substrate can be charged. An external potentiostat holds the potential in a region where non reactions occur at the electrode. In such a way the local etching of the substrate can be performed with high spatial resolution, the latter being a direct function of the pulse duration [86].

In fig. 5.1f a pseudo-electrochemical strategy for local nanostructuring is sketched. The prefix *pseudo* refers to the experimental setup which avoids the use of any reference electrode. Indeed, a tip made of conductive material is held in electrical contact with the surface of a Si wafer through a small amount of an HF aqueous solution. By applying a bias between -2V and 20V for a time between 10ms and 1000s, local formation of SiO_2 occurs. The bias dependence of the rates is a clear indication of the electrochemical nature of the mechanism behind this local oxidation of silicon [76–78].

A very important invention of the last years, at least for electrochemists, has been the Scanning Electrochemical Microscope (SECM). The SECM was developed with the aim to get more insight in the local reactivity of electrodes, giving informations somehow complementary to those available with the STM [88, 89]. Very similar to an STM for its hardware setup, the important difference is that through an SECM-tip flows faradaic current. The tip is usually an ultramicroelectrode (UME) made by a metal wire insulated with a glass shield which also ensures well defined diffusion conditions.

It is evident that the resolution power of the SECM depends on the active area of the tip, that's why Heinze and co-workers developed a method for the construction of UME with an active surface diameter of 20nm called *nanodes* and which allow for a precise determination of reaction rates.

As for the STM, the nanodes are potential tools for local electrochemical modification of surfaces at a nano scale (Fig. 5.1g) which could in the future compete with the STM. But most of the work must be done in order to develop methods for the routinely achieved miniaturizations of the SECM-nanodes [90, 91].

It shouldn't be forgotten that in the electrochemical setup of the STM, the tip is nothing else but a fourth electrode of a cell and then its polarization can be varied at will and if a potential where bulk deposition occurs is chosen, the tip gets loaded with metal from the solution. In a method proposed by Schindler and co-workers, the loaded-tip is approached to the surface and then the metal is re-dissolved in a burst-like way by setting the tip potential shortly at a potential positive of the Nernst value [92–94]. In this way the high local concentration of the cations causes a more positive Nernst potential for the surface region immediately underneath the tip. That results in the local formation of nano-sized cluster as sketched in fig. 5.1h.

The last method, shown in fig. 5.1i, is the topic of the rest of this chapter

5.2 Electrochemical tip-driven nanostructuring

The strategy for a local cluster formation on an electrode surface shown in fig. 5.1i is based on short range forces between tip and substrate which give rise to the so called *jump-to-contact* and was developed through the years in our laboratory. But before describing each step of the method, the key issue of the *jump-to-contact* phenomenon deserves a particular attention.

5.2.1 What is a *jump-to-contact*?

The understanding of the interactions of the tip with the substrate has been, and maybe it still is, one of the key topics in the interpretation of the STM data [95–98].

The extent of the tunneling gap can have strong influences on the image quality. Based on theoretical calculations, Pethica and Sutton described the instability of a system composed by a metal STM-tip brought very close to an ideally flat surface [99]. In such a situation the surfaces on both sides of the tunneling gap jump together causing an irreversible modification of the gap, yielding a lowering of the tunneling barrier. The calculations show that the tunneling distance results shortened in consequence of a displacement of atoms into the gap.

A microscopic description of the phenomenon has been provided by Landman and co-workers [100–103]. Their theoretical calculations reveal the onset

5.2 Electrochemical tip-driven nanostructuring

of an instability as an Au-tip approaches a Ni surface causing a displacement of the Au-atoms filling the space between the two sides of the junction when the tip/surface distance is larger than the lattice equilibrium constants of the two bulk materials. This phenomenon is what is called a *jump-to-contact* (JC).

Let's give some insights of the phenomenon. As shown in 5.2b the JC phenomenon is associated primarily with a sample-induced deformation involving a large atomic displacement (2 Å) occurring in a time span of 1ps.

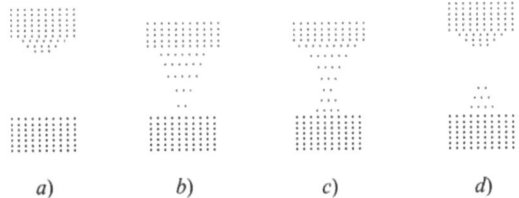

Figure 5.2: *Schematic representation of the jump-to-contact mechanism. a) measuring regime, b) the imbalanced forces in the tunneling gap cause a displacement of the tip, c) JC and formation of a metal neck by diffusion of atoms, d) rupture of the neck by tip retracting.*

The JC phenomenon in metallic systems is driven by the marked tendency of the atoms at the interfacial region to optimize their embedding energies. Retracting the tip after the JC has occurred produces a connective crystalline neck (Fig. 5.2c). The mechanism of elongation of the neck involves atomic structural transformations whereby in each elongation stage atoms in adjacent layers disorder and then form an added layer.

Throughout the process, the neck retains its crystalline structure [101]. It has been also proposed that the connective bridge develops through successive slipping of crystalline layers. Such an elongation mechanism should yield the formation of nano single crystals at the surface [104]. By further retracting the tip, the connective neck breaks and a small portion of material is left behind (Fig. 5.2d). After the discussion above, the JC can be also defined as an inelastic response to the imbalance of the forces acting on both sides of the junction.

On the experimental side, a pioneering work was done by J.W.M. Frenken and co-workers which studied the interaction of an STM-tip with the surface of a single crystal [105–107].

They report that during the imaging of a Pb(111) or Pb(100) sample with a W-tip, a short circuit in the tunneling current is observed. The

CHAPTER 5.Electrochemical nanostructuring with the STM

frequency of the short circuit depends on the Miller index and mainly on the temperature. For values of temperature up to a given threshold it wasn't possible to measure anymore because it was impossible to avoid the short circuit. The results were explained by considering a JC process which brings tip and surface from tunneling regime to ohmic contact [106].

The result of such an instability was the creation of either craters or hillock on the surface. Furthermore, Frenken demonstrated also that the electrostatic forces at the interface don't influence the phenomenon. On the other side, the mechanical properties of the material constituting the two sides of the gap are crucial. Indeed, the atoms of the *softer* material, that is the material with the lower cohesive energy, are displaced into the gap and form the connective neck.

5.2.2 Nanostructuring based on the Jump-to-Contact

During the last decade, in our group a method for the routinely achieved deposition of nano-sized metal clusters based on the JC-mechanism was developed.

In the early studies, the JC was just the explanation for tip instabilities disturbing the measurements, something to be avoided. But in our case the JC is made to happen in a controlled way, being the basic mechanism of the cluster formation *triggered* with the STM tip [71, 72, 108–110].

The cations of the metal to be deposited are in the solution surrounding the tip. The tip in the electrochemical setup of an STM is the fourth electrode of an electrochemical cell, the first being the substrate, held at a potential measured against a reference electrode (RE) by a bipotentiostat which also measures the total current flowing through a counter electrode (CE) (Fig. 5.3a).

Metal deposition onto the tip is achieved by choosing a tip-potential at which bulk deposition occurs. By sending to the z-piezo element of the STM-scanner a voltage pulse, the loaded tip is approached to the surface (Fig. 5.3b) close enough for the activation of the JC from the tip to the surface (Fig. 5.3c). After the JC has occurred, the tip is then retracted and a connective neck is formed (Fig. 5.3d). With a further retreat, the neck gets broken and a small part of material is left on the surface (Fig. 5.3e). The tip is loaded again because of the ongoing metal deposition and the cycle can be repeated at will. During the entire process, the working electrode potential is held in a region where non direct metal deposition from the solution occurs. The process can be performed at khz rates: the generation of 10,000 Cu-clusters on Au(111) is a matter of few minutes [111]. In figure 5.4 a field of Pd-clusters on Au(111) is shown.

5.2 Electrochemical tip-driven nanostructuring

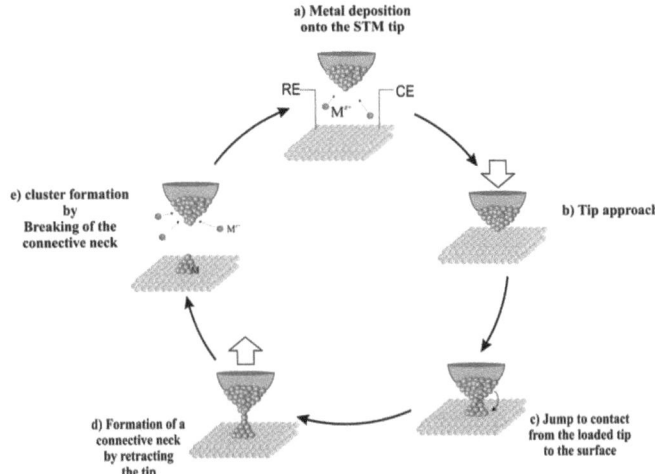

Figure 5.3: *Electrochemical nanostructuring via jump-to-contact mechanism with an STM-tip.*

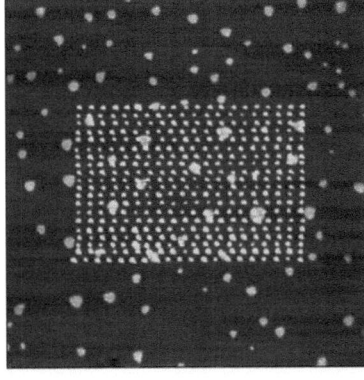

300 nm x 300 nm

Figure 5.4: *Field of 400 Pd clusters on Au(111). E=+670 mV/SCE, electrolyte: H_2SO_4 0.1M+$PdSO_4$ 0.1 mM.*

CHAPTER 5.Electrochemical nanostructuring with the STM

The picture allows already for some important remarks. The potential of the gold electrode during and after the entire nanostructuring process was held at +670 mV/SCE, a value clearly higher of that where a Pd-UPD on Au(111) is deposited [112]. This is already a sign of an high stability of the clusters against the anodic dissolution, at least with respect to the Pd-UPD deposits. The same behaviour has been also observed for Cu-clusters [69, 71]. The topic of the stability of the clusters is very important for several reasons and is extensively treated in section 5.3.

The gold islands stemming from the lift of the thermal reconstruction [113] are visible as bright spots all over the surface and some of them have been decorated with the clusters. Indeed, the nanostructuring occurs without turning off the feed-back control of the STM, so that the tip can still follow the profile of the surface and the presence of steps or defects doesn't disturb the process of nanostructuring. In fig. 5.5 an impressive example of

570nm x 570nm

Figure 5.5: *STM picture of a field of Pd clusters on Au(111) at bunching steps. E=+670 mV/SCE, electrolyte: H_2SO_4 0.1M+$PdSO_4$ 0.1 mM.*

nanostructuring at bunching steps is shown.

It must be said that the tip used for the nano-structuring is the same used as for imaging the surface. This is already a proof that the JC-mechanism is active: a process occurring via contact or crash would destroy the tip, making impossible the successive imaging. Furthermore, by controlling the approach of the tip, the size of the clusters can be varied at will within a given range. As shown in fig. 5.6, 2nm is somehow a limit: a further approach causes the tip crash and no more height variation is observed, but in this case the

5.2 Electrochemical tip-driven nanostructuring

mechanism of the clusters formation is completely different.

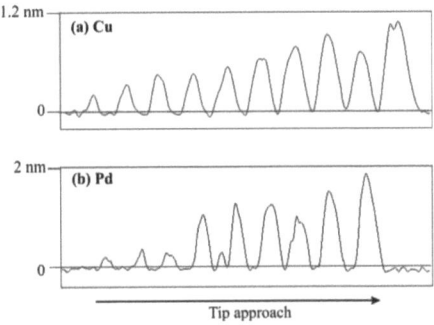

Figure 5.6: *Height variation in the JC tip-driven nanostructuring with the tip approach for a) Cu and b) Pd deposition. a)E= 0 mV/SCE, H_2SO_4 0.1M+$CuSO_4$ 0.1 mM, b) E=+670 mV/SCE, H_2SO_4 0.1M+$PdSO_4$ 0.1 mM.*

In section 5.2.1 the role plaid by the *hardness* of the materials involved in a jump to contact has been stressed, since this is the parameter which determines the direction of the jump: it is actually the *softer* material which jumps onto the harder one. That's why when the tip is loaded with Ni, a material harder than Au, the process produces holes in the surface shown in fig. 5.7: indeed, in this case the surface (Au) atoms jump to the tip, and this is one more evidence that the jump-to-contact mechanism drives the cluster formation.

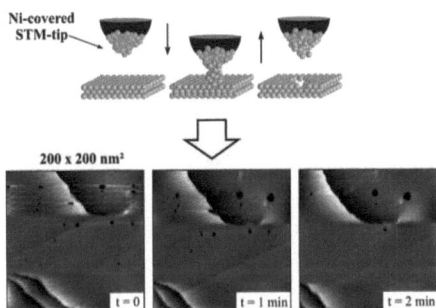

Figure 5.7: *STM picture of holes produced by a Ni-loaded tip (from ref. [111]).*

55

CHAPTER 5. Electrochemical nanostructuring with the STM

In practice, the nanostructuring is achieved by connecting the STM to an external device controlled by a self written software which allows for a fully automated process (for details see the experimental section) which has been tested for the systems listed in table 5.1. The method allows for the

System	Result
Cu/Au(111)	3-D clusters
Cu/Au(100)	3-D clusters
Cu/Au(111)+C_2H_5SH	3-D clusters
Cu/Au(111)+$C_{18}H_{37}SH$	3-D clusters
Pd/Au(111) (*)	3-D clusters
Pd/Au(111)+C_4H_9SH (*)	3-D clusters
Ag/Au(111)	3-D clusters
Pb/Au(111)	3-D clusters
Co/Au(111)	2-D islands
Rh/Au(111)	holes on the surface
Pt/(Au(111)	3-D clusters
Ni/Au(111)	holes on the surfaces
Cu/Ag(111)	2-D islands
Pb/Ag(111)	3-D clusters
Cu/Cu(111)	3-D clusters
Pb/n-Si(111):H	no clusters
Cu/HOPG	no clusters

Table 5.1: *Systems for which the JC mechanism has been demonstrated (from ref. [111] except (*))*

decoration of different kinds of surfaces, although the dynamic properties of the surface can lead to different geometry for the resulting nano deposit [111]. As a further example, the deposition of Cu clusters on an Au(111) surface is shown in fig. 5.8.

Anyway, there is some operative difference between different systems. While the tip can be loaded with Cu in a time of the order of 10ms, the deposition of Pd onto the tip is slower and a loading time of at least 200ms is needed: whereas these different times don't play a role for small arrays, covering an area in the micrometer range makes the difference. The method based on the jump-to-contact for the nanostructuring of surfaces with the STM is not suitable for industrial applications.

It is useful to say something more about the size of the clusters. The Cu clusters are deposited on a Cu-UPD because the electrochemistry of the Cu/Au(111) system allows for a free Au surface only in a range of potentials

5.2 Electrochemical tip-driven nanostructuring

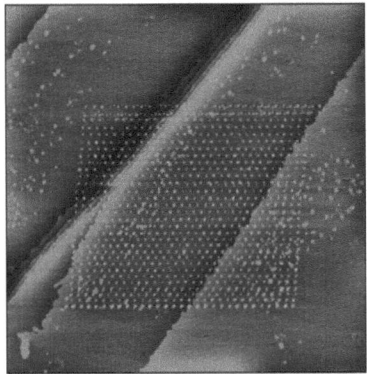

700 nm x 700 mm

Figure 5.8: *STM image of a field of Cu-clusters on Cu-UPD/Au(111). Electrolyte: 0.1M H_2SO_4 0.1M+ $CuSO_4$ 1mM. E=0nV/SCE.*

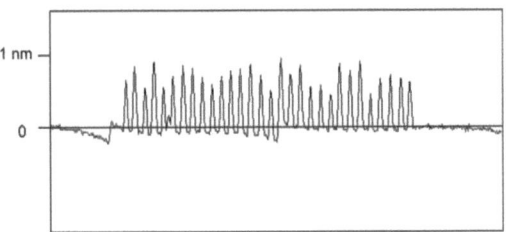

Figure 5.9: *Height profile of Cu-clusters on Au(111). (section measurement of the STM figure 5.8).*

CHAPTER 5.Electrochemical nanostructuring with the STM

where the Cu clusters would be quickly dissolved [114]. This is not the case for Pd which is indeed deposited directly on bare gold. This could be a reason for the more regular height variation for the Cu clusters in comparison with Pd shown in fig. 5.6.

Fig. 5.9 shows the section profile of a single line of clusters of fig. 5.8. It is eye-catching that the size of the clusters involves several Cu monolayers and has an average diameter at the basis of 5nm. However, the repeated conical shape of the clusters led us to suppose that, due to the small dimensions of the deposits, it is actually the profile of the tip which is imaged. That means that their *real* size must be smaller. It is well known that the visualization with STM and AFM may not always produce images which accurately reproduce the size of the sample because of distortions inherent in an imaging process that produces a tip-sample geometric convolution as schematically shown in fig. 5.10 [115, 116]. There are several algorithms which subtract the convolution of the tip profile from the STM image in order to get a more correct image of features present on the sample, but the discussion of this topic goes beyond the aim of this work.

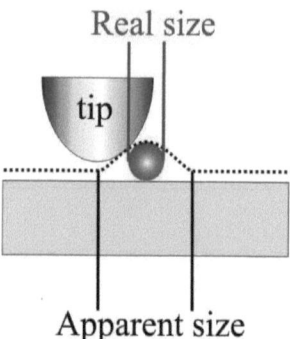

Figure 5.10: *Tip-sample convolution in an STM image.*

The possibility to locally modify an electrode with the STM gives the opportunity to drive or influence the processes occurring at the surface.

It is well known that the metal deposition very often starts at the monoatomic steps and at the defects on the surfaces. If the nano-clusters deposited with the jump-to-contact methods are regarded as defects which can be formed at a chosen position on the surface, the question of a deposition controlled by the creation of nucleation seeds, eventually arranged in special arrays arises as a new fascinating challenge for electrochemistry.

5.2 Electrochemical tip-driven nanostructuring

While the idea seems to be very simple, the first investigations of the influence of the nanoclusters on the metal deposition process appear complicated.

In fig. 5.11 the different steps for the deposition and dissolution of Cu on a Au(111) surface covered with an array of 25 Cu clusters are shown. After the deposition of the Cu clusters on a Cu-UPD (fig. 5.11a), the electrode potential has been set to a more negative potential.

In fig. 5.11b the deposition of Cu at the monoatomic step on the left side and in correspondence of the clusters is evident.

But not all the clusters grow in the same way. By waiting few minutes, it can been seen that there are clusters which don't grow at all (fig. 5.11c), while the others continue to grow along the crystallographic axes of the substrate as demonstrated by the shape of the spots.

After some minutes, the deposition begins also on other places of the surface and in correspondence of the array the deposit still shows empty portions.

This phenomenon can be explained by assuming that once the deposition starts at a precise cluster, its growth removes cations from the diffusion shell of its neighbor.

The question is now what makes the clusters not all the same for the onset of the deposition.

In order to answer this question, the height of each cluster in fig. 5.11a has been measured and is shown in fig. 5.12.

It can been seen that although there is the tendency for the bigger (higher) clusters to act as nucleation sites, this is not always the case since there are some of them which grow regardless of size.

The dissolution of the Cu deposit opens further questions. In fig. 5.11e and 5.11f it can be seen that some cluster didn't react at all but what is more surprising is that the clusters which served as nucleation seeds appear again and they are recognizable as brighter spots (indicated in fig. 5.11f by circles) on the residual Cu layer.

That means that even if those clusters have catalyzed the Cu deposition and the deposit grows all around them, they cannot be considered embedded in the deposit and retain their *character* of clusters as demonstrated by the dissolution of the electrodeposited Cu.

In other words, despite the influence on the modification of the electrode surface, the cluster retain a *memory* of the original array.

A lot of work needs still to be done in order to explain these experimental evidences.

CHAPTER 5. Electrochemical nanostructuring with the STM

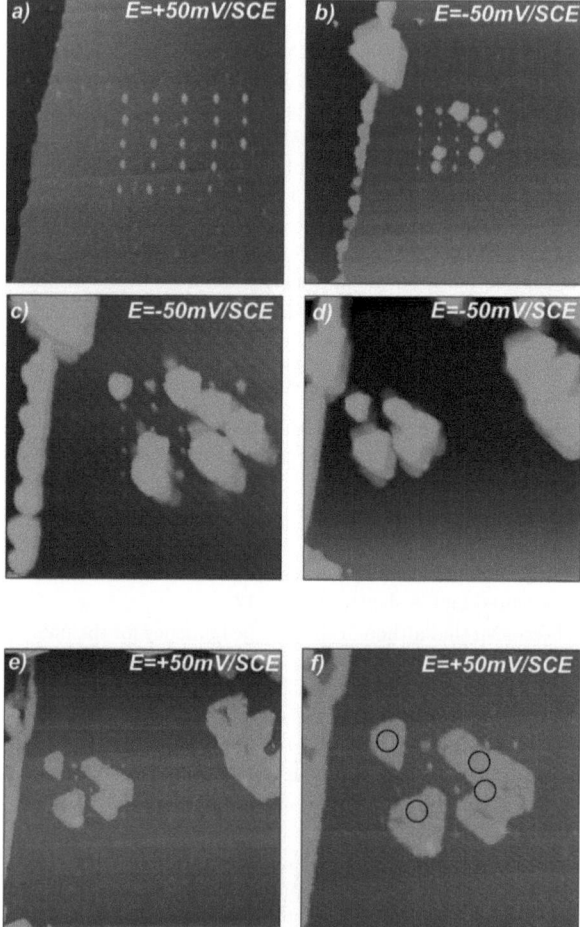

Figure 5.11: *STM images for the different steps of the Cu deposition (b-d) and dissolution (e-f) on Au(111) carrying an array of 25 Cu Clusters (a). Electrolyte: H_2SO_4 0.1M + $CuSO_4$ 1mM*

5.2 Electrochemical tip-driven nanostructuring

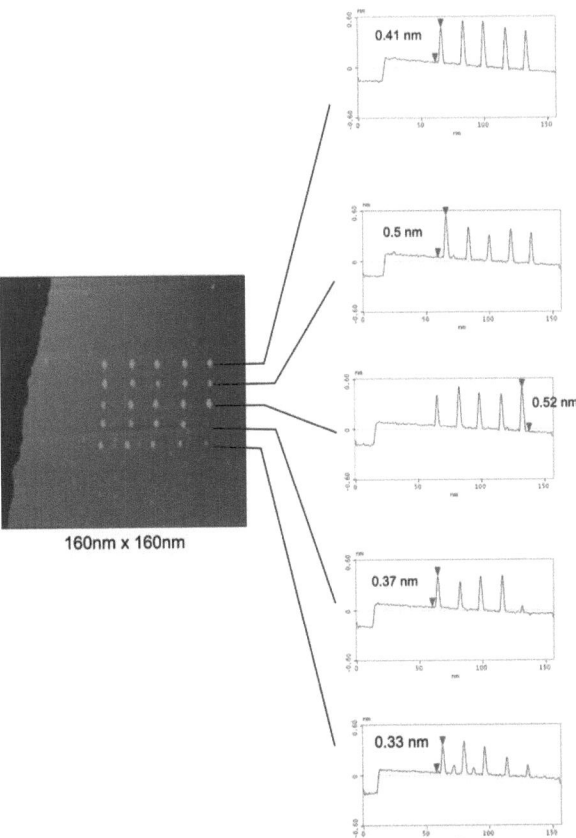

Figure 5.12: *Section analysis of the cluster array shown in fig. 5.11a*

CHAPTER 5.Electrochemical nanostructuring with the STM

5.3 Pd clusters on Au(111)

The jump-to-contact method for the *in-situ* nanostructuring of the surface has been extensively used for the formation of Pd clusters on a Au(111) surface.

In order to better understand the behaviour of the Pd nano-clusters and also to appreciate the potentialities of the electrochemical nanostructuring which are the main topics of this chapter, the Pd deposition on Au(111) is first shortly reviewed.

5.3.1 Electrodeposition of Pd: an overview

The deposition of Pd on Au(111) has gained a considerable attention during the last years. The peculiarities of thin Pd layers made this system to be chosen as model for the study of special electrocatalytical reactions.Among the other systems, a special place deserve the hydrogen adsorption and evolution [123], formic acid oxidation [124, 125] and the oxidation of CO adlayers [126].

This special reactivity of Pd thin layers is due to their structure. Experimental works [112, 123, 127–129] and theoretical calculations [130–132] reveal that the first and second layer of Pd on Au(111) are *pseudomorphic* with the substrate, that means that the adlayers assume a larger lattice constant respect to their bulk phase and that's the reason for an increased reactivity in comparison with bulk Pd crystals [130, 132].

It is known that Pd alloys with Au(111) at room temperature under UHV conditions [133], but no alloy formation has been detected in the electrodeposition from a solution containing Pd cations onto Au(111) electrode. That's exactly the reason why the deposition of Pd clusters on Au(111) has been chosen for an extended exploration of the nanostructuring method based on the JC.

In fig. 5.13, a cyclic voltammogramm for the deposition of Pd on Au(111) is shown. As it has been proven with the STM, the peak at +0.56V is due to the formation of a pseudomorphic Pd-UPD layer [112]. At this potential, the pseudomorphic growth is observed only at the monoatomic steps without the formation of a complete layer. The peak at 0.45 V is associated with the growth of a second monolayer which overlaps with the further bulk depostion.

By decreasing the potential, the bulk deposition starts a the bunching steps sites and in the first stages the deposit retains its pseudomorphic character.

Pd layers have been deposited also from solutions containing Cl^-. In this case the deposition occurs via the reductive discharge of the adsorbed Pd-chloro-complexes at 0.6 V [128].

5.3 Pd clusters on Au(111)

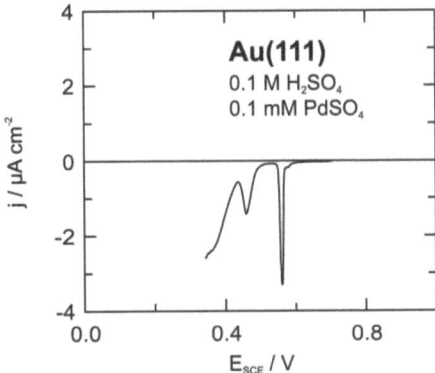

Figure 5.13: *Cyclic voltammogramm of Au(111) in 0.1M H_2SO_4+0.1mM $PdSO_4$. (Courtesy of Mrs. J. Tang)*

The absence of complexes or other species which could trap the Pd cations allows for an enhanced mobility of the ions at the surface in the chloride free solution and this explains the growth of 2-dimensional round terraces observed with the STM for Pd deposition from solutions where only H_2SO_4 is used as support electrolyte [112]. On the other hand, the chloro-complexes are somehow blocked at the surface and the growth follows the geometry of the underlying substrate, as demonstrated by the characteristic Pd triangles in the first stages of deposition observed with the STM [128].

A different stability of Pd layers deposited from different solution has been observed. While in Cl^- containing electrolyte the dissolution of Pd layers is enhanced by the formation of Pd chloro-complexes, in H_2SO_4 this process is extremely slow because a rather stable oxide is already formed around 0.7 V. Only by repeating for several times an oxidations/reduction cycle a gradual dissolution takes place [112].

The next sections deal with Pd clusters deposited from a Cl^--free, H_2SO_4 solution.

5.3.2 Pd clusters on Au(111)

The deposition of Pd on Au single crystals occupied a significant part of the publication in electrochemistry of the last recent years.

Not only the interesting topic of the growth of pseudomorphic layers attracted the attention of many researchers, but also, and maybe manly, the

CHAPTER 5. Electrochemical nanostructuring with the STM

catalytical properties of such thin layers which are very stable and can be obtained at a very high degree of purity.

These are the reasons why the modification of an Au(111) surface with Pd clusters became the main application of the jump-to-contact based method, namely the deposition in a controlled way of nano-sized particles of an active material on an inert substrate.

As an example, fig. 5.14 shows the result achieved in our laboratory. The deposition of these clusters was achieved by keeping the potential of the

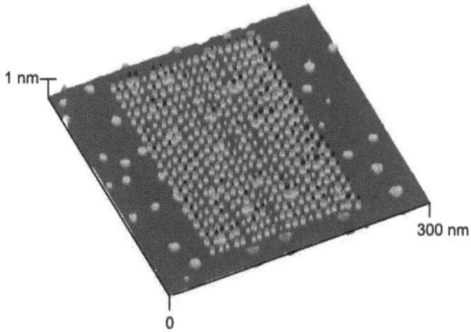

Figure 5.14: *STM image of an array of 400 Pd clusters on Au(111). E=670 mV/SCE, electrolyte: 0.1M H_2SO_4 +0.1mM $PdSO_4$.*

gold electrode in a region slightly positive of the UPD value. It must be said that the the formation of Pd clusters was possible also at very positive potentials, where the Au surface is positively charged and the Pd layers are dissolved. This is a further sign of the exceptionally high stability of these nano-deposits.

Beside the electrochemical conditions, in the experimental formation of the clusters there are other parameters that must be optimized.

It was observed that the tip must be kept in the vicinity of the surface for a time not less than 4ms in order to activate the JC. This *pulse duration* measures the duration of the voltage-pulse sent to the piezo crystal of the STM head in order to approach the tip to the surface. As already mentioned, during the entire process the feed-back loop of the STM is not switched off

5.3 Pd clusters on Au(111)

and in correspondence of the JC the controller of the STM reacts to the high current by withdrawing the tip and trying to adjust the current to the set point value. This fact is thought to influence the observed time needed for the JC to occur.

An important parameter is the *waiting time*, namely the time the tip spends away from the surface in order to get loaded. In the case of Pd, due the very slow deposition rate onto the Pt/Ir tip, the waiting time must be at least 150 ms. A shorter waiting time for Cu (10-20 ms) allows for a faster process in comparison with Pd.

Both the pulse duration and the waiting time determine the frequency of the nanostructuring. Significant Pd cluster formation is usually achieved at a rate of 6 clusters per second. The clusters shown in fig. 5.15 were deposited under these experimental conditions.

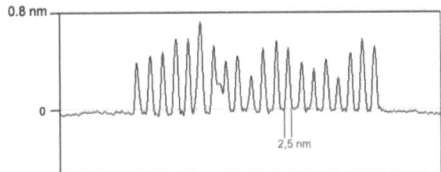

Figure 5.15: *Lateral section of a Pd-clusters array on Au(111).E=670 mV/SCE, electrolyte: 0.1M H_2SO_4+0.1mM $PdSO_4$.*

The relevant topic of the apparent size of small features imaged with the STM has been already discussed in section 5.2. A value of 2.5 nm for the diameter at the basis of a Pd cluster shown in fig 5.15 has been roughly calculated via deconvolution of the tip geometry which leads to a doubled value in the experimental profile [115, 116]. The height is correctly reproduced.

That implies that *in reality* the clusters look like islands composed by nearly 3 Pd-monolayers. By approximating their shape to a cylinder with a diameter of 2.5nm and height of 0.6 nm (fig. 5.15) and considering an atomic radius for Pd of 0.138 nm, it can be calculated that the clusters contain a number of atoms less than 100.

Unfortunately, informations on their structure are scarce and at the present extremely difficult to get.

Based on theoretical models, some authors have argued that the JC mechanism leads to a perfect mono-crystallinity of the clusters [104]. On the other hands, a theoretical calculation of the structure of very small (3 to 10 atoms) Pd clusters on an Au(111) surface geometry yields a reduced interatomic distance in the clusters with respect to the bulk Pd lattice constant: if it were true also for bigger, real deposits, the cluster should have a compressed structure instead of an expanded, pseudomorphic geometry [134]. Experimental informations other than STM are still lacking.

Some questions arose about the composition of the clusters. The formation of an alloy has been assumed for the explanation of the high observed stability [135, 136]. The jump-to-contact mechanism for the nanostructuring doesn't involve a mixture of atoms as it were for a nano-identation where the eventual alloy formation is the result of a tip crash. On the other hand, there are experimental evidences that exclude the alloy formation. Engelmann already reported that by dissolving Cu nanoclusters no defects or feature typical for local alloy formation are observed [111]. The case of Pd cluster is very similar to Cu and it is treated in the next section. Anyway, the alloy formation during and after the JC can be excluded although the characterization of the clusters remains still an open problem.

Very problematic is also the study of the behaviour of such nano-sized systems. In order to make possible electrochemical measurements, a large area of the surface must be decorated with nano-clusters. As an example, the nanostructuring over a micrometer-scale region of fig. 5.16 takes few hours, but remain still a far too small area for the traditional electrochemical methods like cyclic voltammetry. Some different, local approach must be attempted.

In fig 5.17 an experiment of Pd deposition on a Au(111) surface previously modified with Pd-nano clusters is shown. It can be noticed that the clusters act like nucleation seeds, making possible to localize the onset of the growth of a metal deposit.

The JC method allows also for the deposition of nano clusters on modified metal surfaces. In fig. 5.18, Pd clusters fields deposited onto a Au(111) covered with an organic self assembled monolayer are shown.

5.3.3 The stability of the clusters

The stability of the nano-clusters generated with the jump-to-contact method is a fascinating issue, a topic which puts questions but does not yet allow for unambiguous answers!

The tip-generated clusters show an unusually high resistance against the anodic dissolution. It has been already mentioned that a possible explanation proposed in the literature is the alloy formation. This possibility has been directly checked.

In fig. 5.19 the cluster field a) has been deposited on bare gold. After that a Pd-UPD has been growth and the fields b) have been deposited onto

5.3 Pd clusters on Au(111)

2.5µm x 2.5µm

Figure 5.16: *STM image of an Au(111) surface covered with Pd-cluster fields each containing 2,500 clusters. E=670 mV/SCE, electrolyte: 0.1M H_2SO_4+0.1mM $PdSO_4$.*

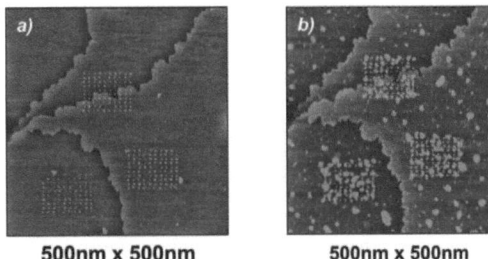

500nm x 500nm   500nm x 500nm

Figure 5.17: *STM image of Pd nanoclusters on Au(111). Electrolyte: 0.1M H_2SO_4+1mM $PdSO_4$+5mM HCl. a) E=650 mV/SCE, b) E=440 mV/SCE.*

CHAPTER 5.Electrochemical nanostructuring with the STM

Figure 5.18: *STM image of Pd clusters deposited on a C_4H_9SH SAM covered Au(111) surface. Electrolyte: 0.1M H_2SO_4+ 0.1mM $PdSO_4$.*

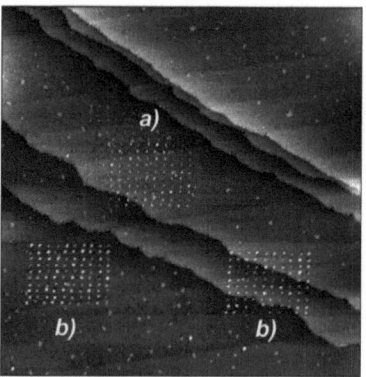

Figure 5.19: *STM image of Pd clusters deposited a) on bare gold and b) on a Pd UPD. Electrolyte: 0.1M H_2SO_4+0.1mM $PdSO_4$+5mM HCl.*

5.3 Pd clusters on Au(111)

the monolayer. This is the reason for the different brightness (height) of the clusters of the different fields.

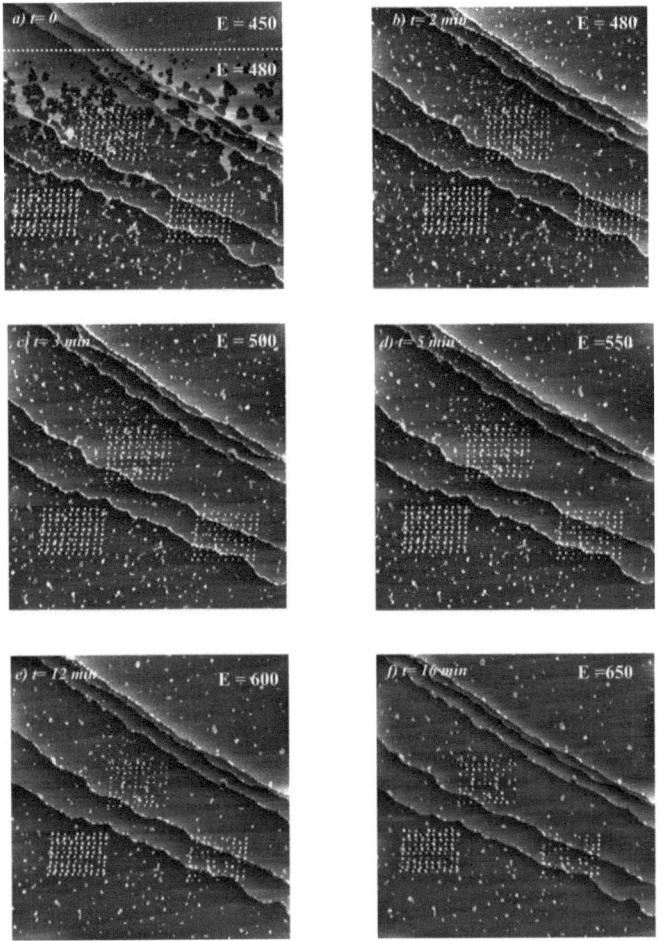

Figure 5.20: *Stability of Pd clusters on Au(111) versus anodic dissolution. Surface area 500nm x 500nm. Electrolyte:0.1M H_2SO_4+0.1mM $PdSO_4$+5mM HCl.*

CHAPTER 5.Electrochemical nanostructuring with the STM

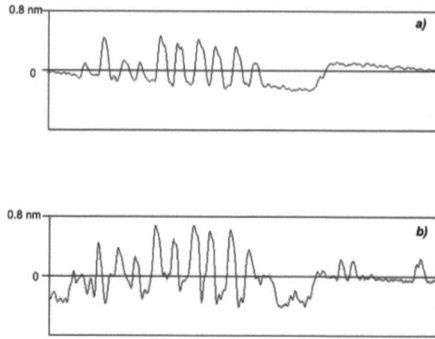

Figure 5.21: *Comparison of the section height of Pd Clusters on Au(111) at two different potentials. Electrolyte:0.1M H_2SO_4+0.1mM $PdSO_4$+5mM HCl. a) Clusters on a Pd-UPD, E= 450 mV/SCE, b) Height of the cluster after the dissolution of the Pd-UPD, E=480 mV/SCE.*

It is well known that Pd adlayers on Au(111) do not alloy under electrochemical condition [112, 128]. In this way, if the doubt for an eventual alloy formation still persists for the field a), it this removed, or at least strongly minimized for the two fields b).

The electrode potential is hence set to a value where Pd dissolution occurs. The behaviour of the system at different times is shown in fig. 5.20.

The direct comparison of fig. 5.20a with fig. 5.20b demonstrates that the Pd clusters remain at the surface also when the UPD is dissolved. The measurements of the profiles before and after the Pd-UPD dissolution reported in fig. 5.21 show an apparently increased height for the clusters. This is off course the effect of lowering the background zero line. This experimental evidence demonstrates that during the dissolution of the UPD no dissolution occurs on top of the clusters, which indeed appear higher! Only several minutes after the UPD has been removed the clusters start slowly to dissolve (fig. 5.20 d-f). A direct comparison of the cluster profiles at the beginning and a the end of the process is provided in fig. 5.22.

In fig. 5.20f, it can been noticed that at the location of the missing, dissolved clusters, no defects or spots are detected with the STM, neither for the field deposited directly on bare gold, nor for those deposited on the UPD: there is no relevant difference in the behaviour of the two kinds of fields and

5.3 Pd clusters on Au(111)

thus the alloy formation can be exclude.

Several different attempts have been reported in the literature in order to give an account for the stability of not only Pd clusters.

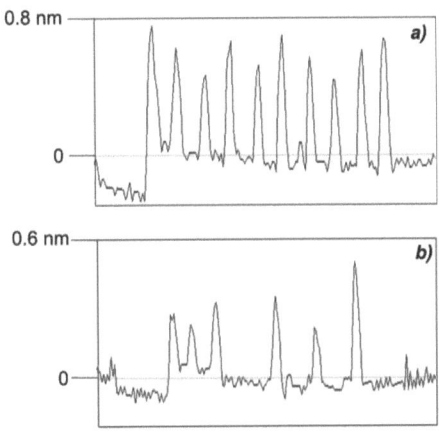

Figure 5.22: *Variation of the height of Pd clusters on Au(111) with the potential. Electrolyte: 0.1M H_2SO_4+0.1mM $PdSO_4$+5mM HCl.* a) E= 480 mV/SCE, b) Dissolution of the clusters, E= 650 mV/SCE.

Lorenz and co-workers tried to give a thermodynamical description of the stability of *Low Dimension Systems* (LDS) like zero-dimensional nanoclusters, 1-dimensional chains or nanowires and 2-dimensional expanded or condensed monolayers [137]. His model is based on the introduction of an activity term which strongly depends on the dimensionality of the system and defines a Nernst-like equation for the equilibrium potential of the LDS, the lower the dimensionality, the more positive is the Nernst-like potential associated.

But this seems to be an *a posteriori* explanation, and since the LDS are made by a small number of atoms, the *black box* thermodynamical approach is not appropriated. An atomistic, extra-thermodynamic treatment is necessary.

CHAPTER 5.Electrochemical nanostructuring with the STM

As already mentioned, Pd-adlyers on Au(111) have an expanded, pseudomorphic structure [127]. Recent theoretical calculations of the structure predict a reduced interatomic distance for small clusters [134]. This atomic compression is associated, and partially caused, with a strong coupling of the clusters with the underlying substrate which leads to an increased stability [132]. Nevertheless, these theoretical clusters are composed only by a too small number of atoms compared with the real ones, although these results are a clear indication that *small behaves different*.

Due to the reduced size, some quantum effect cannot be excluded.

The electronic confinement in metals leads to a discrete energy levels spectrum with a spacing given by [138]:

$$\Delta E_{conf} = (2/3)\frac{E_F}{N} \tag{5.1}$$

where E_F is the Fermi energy and N the number of atoms. By considering $E_F \approx 10eV$ and $N = 100$ [138], an energy separation of 0.1 eV is expected.

This is a very rough estimation, but it gives already the idea of how important are the quantum effects for electrochemical measurements since a energy spacing of 0.1 eV corresponds to a potential shift of 100 mV in, e.g. the dissolution process [69].

If the above considerations are able to give an account for the behaviour of the clusters on the basis of the quantum theory, it should not be forgotten that the clusters are not isolated and it is not yet clear what's the effect of the supporting surface on their electronic structure.

The stability of the electrochemically deposited nano-clusters is far to be completely understood. The actual and sometimes contradictory explanations need a lot of work on the experimental as well as on the theoretical side in order to be overcome. A more detailed quantum analysis is needed together with a refinement of the theories and at the moment it seems not a simple task [139].

Nonetheless, the stability of Pd clusters is an experimental evidence which gives us the possibility to prepare samples suitable for different measurements which require the removal of the cells from the STM by losing the potential control.

In order to get an idea about the life time of the nano-modified surfaces, an Au(111) sample carrying Pd clusters has been kept for 10 days in pure water and without potential control.

The STM picture shown in fig. 5.23b demonstrates that after such a long time no sign of dissolution or degradation like lateral diffusion or relaxation to 2-dimensional islands can be detected.

5.4 First investigation of Pd cluster reactivity with the SECM

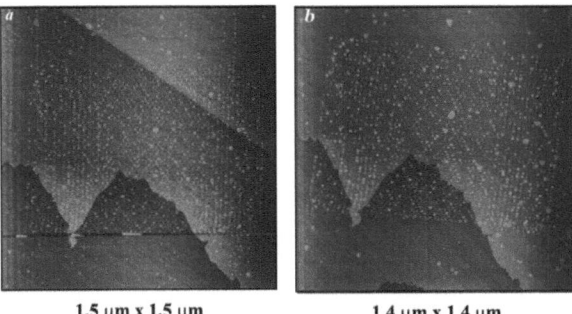

1.5 µm x 1.5 µm   1.4 µm x 1.4 µm

Figure 5.23: a) Pd clusters on Au(111). E=660 mV/SCE, electrolyte: 0.1M H_2SO_4+1mM $PdSO_4$. b) STM image of the clusters after 10 days kept in pure water. E=660 mV, electrolyte: 0.1M H_2SO_4

This allows to prepare samples of Au(111) crystals modified with Pd nanocluster which survive for a long time. The samples can then be transported by train to another Institute, where they can be investigated. e.g. with the Scanning Electrochemical Microscope.

5.4 First investigation of Pd cluster reactivity with the SECM

(All the methods and the experiments discussed in this section are the result of a collaboration with Elmar Laubender of the Department of Physical Chemistry I-FMF directed by Prof. Jürgen Heinze at the University of Freiburg, Germany. Technical description of experiments and devices can be found in ref. [117])

As it will be discussed in chapter 6, some evidences of the electroacatlytical activity of the Pd nano-clusters have been observed with the tunneling spectroscopy measurements.
Although this kind of measurements are able to reveal localized process occurring at the surface, a quantitative analysis of the results is not possible simply because shape and size of the active surface of an STM tip are unknown.

Local measurements under well defined electrochemical conditions can be performed with the Scanning Electrochemical Microscope (SECM).

CHAPTER 5. Electrochemical nanostructuring with the STM

The hardware set up of a SECM is very similar to the STM. A tip-probe is moved upon a surface with the help of micrometric motors as well as piezo elements for the three spatial directions. The tip and surface potentials are measured against a reference electrode (RE) by a bipotentiostat. The circuit is closed with a counter electrode (CE) (fig. 5.24). In a SECM, at the tip flows faradaic current and that makes the *conceptual* difference with the STM.

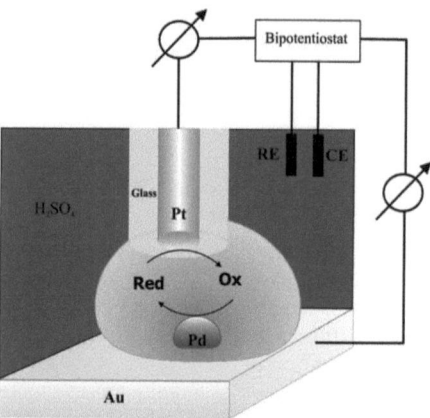

Figure 5.24: *Schematic representation of principles and experimental set up of a SECM*

The invention of the SECM opened a new era for electrochemistry because can probe a localized region of the electrode giving informations for different kind of reactions like electron, ion and molecules transfers at different places of the surface. Furthermore, the SECM allows for measurements under well defined electrochemical conditions also at liquid/liquid and liquid/air interfaces [118, 119].

In a SECM a *ultramicroelectrode* (UME) is employed as tip probe. UME are electrodes with a diameter of the active surface $\leq 20\mu m$. When the diameter is $\leq 1\mu m$, they are called also *nanodes*. This kind of electrodes deserves a special attention because their behaviour is determined by the reduced size and has important implication for the understanding of other electrochemical processes occurring at the nano-scale.

The advantage to use UME in the SECM is due to their unusual properties characterized by an extremely reduced ohmic resistance and by the fact that UME reach a steady-state in a very short time respect to the macroelectrodes.

5.4 First investigation of Pd cluster reactivity with the SECM

It is evident that the beahviour of the UME is *qualitatively* different of that of normal planar electrodes, and this is a direct effect of their diminished size [120].

For a planar macroelectrode in contact with a solution containing an electroactive substance of concentration c^*, by applying a potential sufficient to completely consume the species, the Cottrel's theory predicts a decay of the current i which obeys the following equation:

$$\frac{it^{1/2}}{nFAD^{1/2}c^*} = \frac{1}{\pi^{1/2}} \qquad (5.2)$$

Where D is the diffusion coefficient. The decay of the current with $t^{-1/2}$ predicted by Eq. 5.2 derives from solving the Fick's equations for the case of planar diffusion. Due to the size of the active surface of a macro-electrode, edge effects and lateral diffusion can be neglected (fig. 5.25 left)

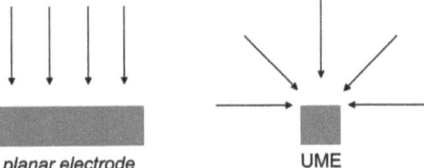

Figure 5.25: *Diffusion geometry for planar and UME electrode*

By reducing the size, the lateral diffusion becomes important, the smaller the electrode, the greater the lateral diffusion contribution (5.25 right). From a qualitative point of view, the hemispherical diffusion feeds the electrode more efficiently because of an increased solid angle of the diffusion layer in front of the electrode. In order to calculate the expression of the current at the UME, the solution of the Fick's equations for the *hemispherical* diffusion yields the following expression:

$$\frac{it^{1/2}}{nFAD^{1/2}c^*} = \frac{1}{\pi^{1/2}}\left(1 + \pi^{1/2}\left(\frac{Dt}{r_0^2}\right)^{1/2}\right) \qquad (5.3)$$

Eq. 5.3 predicts that after a short time, the first term of the right-hand side is dominating and the equation reduces to the Cottrel expression Eq. 5.2. For long times, the second term dominates resulting in a current independent from the time.

This lapse time for shifting from the planar to the hemispherical diffusion is determined by the radius r_0 of the active surface, and in agreement with Eq. 5.3, the smaller the size, the shorter the time.

CHAPTER 5. Electrochemical nanostructuring with the STM

As it can be expected, the UME reaches in a very short time the limit value of the diffusion controlled current. For an hemispherical diffusion this limiting current is given by [120]:

$$i_d = 2\pi r_0 n F D c^* \qquad (5.4)$$

The limiting current of Eq. 5.4 doesn't depends on the area of the electrode as for planar electrode, but only on the radius r_0. The important consequence is that the current density increases by reducing r_0, whereas with macroelectrodes it is usually independent on the surface area.

Since the reason of all these effects is the reduced size, the description of the behaviour of the UME and of their diffusion mechanism can also be useful for the explanation of nano-scale electrochemical processes, included eventually the activity of the clusters which, due their size, could develop an hemispherical diffusion layer leading to an increased activity.

For all these reasons the UME are not replaceable tools for the SECM, in order to locally resolve the electrochemical behaviour of an electrode surface.

The SECM can be used in different modes, dependently on what kind of informations are needed.

Usually, the *feedback mode* is used for getting a morphological image of the surface. The principles of the feedback modes are depicted in fig. 5.26.

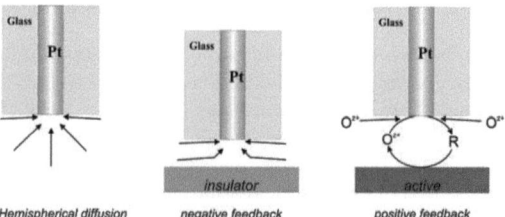

Figure 5.26: *Schematic representation of the feedback modes of an SECM.*

When the tip is given a potential sufficient to consume an electroactive substance, the hemispherical diffusion determines the current at the UME (fig. 5.26 left). When the tip is approached to an inert substrate, the diffusion is disturbed and the electrode can't be fed efficiently. That results in a current decreasing with the distance: the closer the tip, the lower the current. That's what is called a *negative* feedback (5.26 middle). Due to the variation of the the current with the distance, the negative feedback mode allows also for a precise measure of the tip/sample distance.

5.4 First investigation of Pd cluster reactivity with the SECM

When the substrate is active and able to produce again the species consumed at the tip, the current at the UME increases because the depletion layer is continuously reloaded by the substrate. This is a *positive* feedback (fig. 5.26 right).

In our experiments, the feedback mode was used in order to get a morphological image of the substrate. The substrate was a bead Au single crystal obtained with the procedure described by Cavilier [121, 122].

These crystals have the shape of a drop of some millimeter showing very flat Au(111) facets. A photograph is shown in fig. 5.27 left. The correspon-

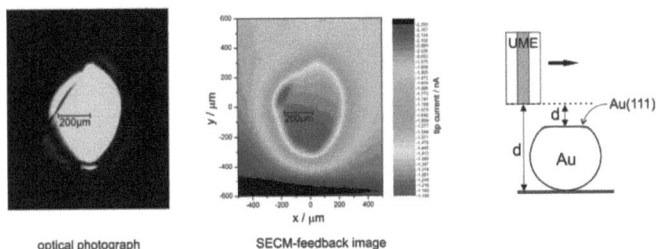

optical photograph   SECM-feedback image

Figure 5.27: *Optical photograph and Feedback-SECM image of a Au(111) facet in H_2SO_4 0.1M. Middle: E_{tip}=-250mV/SCE. Tip/surface distance= 5μm.*

dent SECM image was obtained in negative feedback mode by producing hydrogen a the tip. By starting the measurements in a region far away from the facet, the tip records just the current due to hemispherical diffusion lmechanism of the UME (blue zone in fig. 5.27 middle). By approaching the surface (I remind that the all sample resembles a sphere), the diffusion of protons or hydronium ions toward the tip is disturbed, resulting in a reduced (less negative) current which has a minimum just on the Au(111) facette where the tip is at its closest approach to the surface (red color in fig. 5.27 middle). This is a typical example of the use of a negative feedback.

It is interesting to note that in fig. 5.27 also the defect (a scrape on the left up side) is reproduced in the SECM. Anyway, the resolution of the SECM, although can be improved by minimizing the active area of the UME, can't be compared to that of the STM.

Electrochemical information with the SECM are normally achieved with the *Generator-collector mode*. In this case the products of an electrochemical reaction occurring a the substrate (the generator) diffuses to the tip where there are detected by giving the tip a potential able to oxidize or reduce them. Differently from the feedback mode, in this case a reaction occurs at

CHAPTER 5. Electrochemical nanostructuring with the STM

the tip only when the species originating a the substrate reach the UME. The GC mode is shown in fig. 5.28.

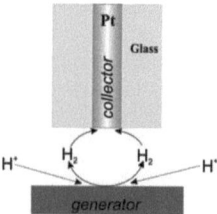

Figure 5.28: *Principle of the Generator Collector mode of a SECM*

The generator collector mode can be used also in the opposite direction by using the tip as generator and measure the variation of the current at the substrate which works as collector. This mode is called the tip generator/substrate collector mode (TG-SC) [118, 119].

Since we were interested in the activity of Pd clusters with respect to the hydrogen evolution, a Clavilier-like crystal carrying a Pd-nanocluster field like that shown in fig. 5.16 was used as generator.

The sample was given different potentials and the relative currents recorded at the tip while scanning the electrode surface are shown in fig. 5.29. In the figure, the feedback measurement is also shown (dotted line) in order to relate the activity to the morphology of the substrate.

The results shown in fig. 5.29 are quite surprising. It is very evident that the signal coming from the edges of the Au(111) facet is quite strong and the current decreases on top of the facet (represented by the flat upper part of the feedback curve), where an higher current is expected because of the presence of the Pd clusters.

Due to the gradual change of the surface orientation at the border of the Au(111) facet, the surface has an high density of steps. It is also well known that bunching steps sites are more reactive than flat surfaces, but the effect measured in fig. 5.29 is extremely high to be explained in this way.

As counter proof, the measurements were repeated for the same single crystal after it was electropolished (see experimental section). With the electropolishing the Pd clusters are completely removed together with all contaminants eventually present on the surface. The reactivity of the bare gold surface was then measured with the SECM in generator collector mode with spatial resolution. The results are shown in fig. 5.30.

It is evident, that also by assuming an higher reactivity at the border of the Au(111) facet, the effects shown in fig. 5.29 are not reproduced and the

5.4 First investigation of Pd cluster reactivity with the SECM

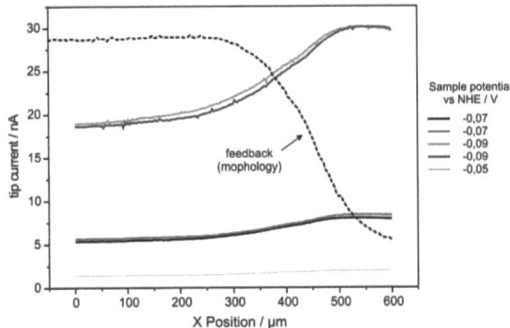

Figure 5.29: *SECM-Generator collector measurement for a Au(111) surface carrying fields of Pd-clusters in H_2SO_4 0.1M. E_{tip}=100mV/SCE, Tip/surface distance= $5\mu m$. Dotted line: feedback measurement*

reactivity of the substrate follows the morphology (feedback). For some reason, the surface measured in fig. 5.29 has an higher reactivity independently of the presence of the clusters.

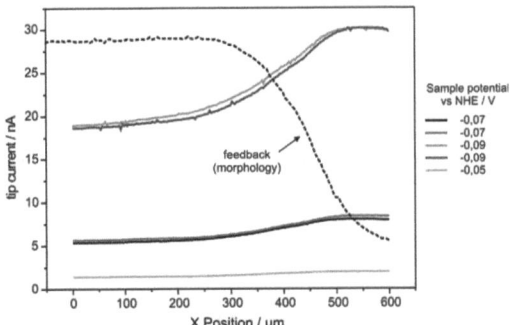

Figure 5.30: *SECM-Generator collector measurements on Au (111) in H_2SO_4 0.1M. E_{tip}=100mV/SCE, Tip/substrate distance= $5\mu m$. The sample potentials on the right side are quoted vs. Hg/Hg_2SO_4 (+0.44V vs. SCE) reference electrode.*

The measurements were repeated for the same crystal after a Pd monolayer was deposited on it. It is well known from electrochemical measurements that these Pd thin films have an enhanced electrocatalytical activity

CHAPTER 5. Electrochemical nanostructuring with the STM

which was confirmed by the SECM measurements shown in fig. 5.31. Quite surprisingly the measurements are very similar to that reported in fig. 5.29 and that may yield the conclusion that all over the surface carrying the clusters there must be Pd.

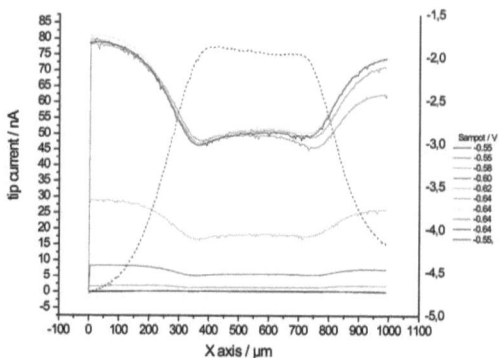

Figure 5.31: *Generator collector SECM measurements on Pd layers on Au(111) in H_2SO_4 0.1M. Dotted line: feed-back*

A further proof of such unexpected activity was directly given by generator-collector measurements on a flat Au crystal where only one well defined portion separated by the rest of the surface with a sealing O-ring, came in contact with a Pd-containing solution. By scanning upon the different regions of the surface, a surprising high activity were detected on the region which was in contact with the solution although the entire electrode was given a potential where no Pd deposition occurs. These results are shown in fig. 5.32.

After this phenomenon was revealed, in our lab the understanding of this *unusual* Pd deposition became a field of intense research. Surfaces which just come in contact with a solution containing Pd ions show an increased surface activity and *ex-situ* measurements revealed a large amount of Pd on the sample while no Pd deposition has been seen with the STM (Private communications with Dr. L. Kibler). Where is exactly this Pd? And how does it reach the surface?

As often happens in science, while we were trying to study the reactivity of the Pd clusters, a new and unexpected phenomenon appeared. Unfortunately, while these experimental evidences stimulated the research in the

5.4 First investigation of Pd cluster reactivity with the SECM

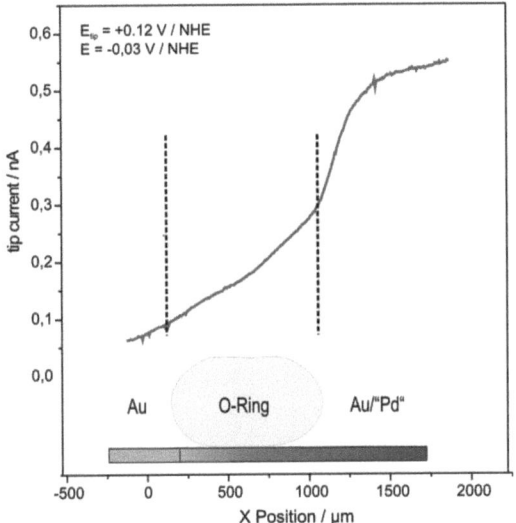

Figure 5.32: *Generator collector SECM measurements upon different regions of an a Au(111) surface in H_2SO_4 0.1 M*

CHAPTER 5. Electrochemical nanostructuring with the STM

field of metal deposition, the too high currents produced by this effect obscure the signal eventually coming from the clusters which is expected to be of the order of few nano amperes.

In order to avoid the contact of the entire surface with the Pd-containing solution during the nanostructuring, we attempted to insulate the Au(111) with an organic self assembled monolayer. The cluster were deposited directly on the modified surface and samples like that shown in fig. 5.18 were investigated.

No activity was observed. Indeed, the cluster formation occurs by locally removing the organic molecules underneath the tip which could re-adsorb at the clusters blocking every electrocatalytical activity.

Moreover, working with SAM is problematic also for the SECM. During the measurements, a strange behaviour of the UME could be observed, maybe due to adsorption of organic molecules at the active surface.

While a new protocol for the preparation of the sample must be developed, the determinations of the electrocatalytical activity of the Pd clusters generated with the jump-to-contact method remains still an open problem.

Chapter 6

Local Analysis of the Electrified Interface

6.1 Introduction

While the imaging of electrode surfaces *in-situ* with an STM, even with atomic-scale resolution is by now a well established technique, in recent years the use of the tunneling spectroscopy is emerging as a method capable to yield valuable structure informations normal to the surface, which otherwise are difficult to obtain.

This is not at least caused by experimental problems arising from the rather limited potential window set by the decomposition potential of aqueous solutions and by the fact that the tunnel voltage between tip and sample not only governs the tunneling process, but also and foremost the electrochemistry at both tip and sample.

Anyway, a direct correlation of the results given by the tunneling spectroscopy with the spatial distribution of the double-layer constituents normal to the surface is almost impossible without theoretical support.

On the other hand, such information may be considered the *missing link* in the double-layer studies: whereas there is a wealth of data on the lateral distribution of ions on single crystal noble metal electrodes, informations on the normal distribution are scarce.

Not only a tool for the investigation of the structure of the interface, the tunneling spectroscopy allows for a deeper understanding of the electron transfer at the electrified interface because it gives a direct access to the parameters governing its fundamental tunneling mechanism which in turn determines the current flowing through the interface (see section 2.2). This is an issue of crucial importance not only for electrochemists.

CHAPTER 6.Local Analysis of the Electrified Interface

6.2 Some particular aspect of the *in-situ* STM

A general result of the tunneling spectroscopy is that the measured value of the tunneling barrier height ϕ_T calculated with the expression (see section 2.3):

$$\phi_T(s) = \frac{\hbar^2}{8m}\left(\frac{\partial ln I_T}{\partial s}\right)^2 \qquad (6.1)$$

is always lower than the average of the work functions of the two metals on either side of the gap [153–155]. Indeed, at usual gap-width in an STM, the electron encounters a lowered barrier height, a phenomenon essentially due to exchange and correlation effects, as discussed in section 2.1.

In the *in-situ* experiments the tunneling barrier is further reduced because of the interactions of the tunneling electron with the solvent inside the gap, although theoretical calculations prove that only the optical component of the dielectric modes of the solvent are relevant, since the tunneling is an extremely quick process[156, 157].

Besides the electronic factor, the structural properties of the liquid part of the interface are also important.

It has been reported that at the electrified interfaces the tunnel current decay is not strictly exponential, but shows a more complicated dependence on the distance and on the electrode potential just because the electrolyte assumes a precise structure as response to the electric field created at the interface by the applied potential (section 1.3) [140–152].

But the role of the solvent is much more complex: other aspects have attracted the attention of the electrochemists and other possibilities have been proposed in order to better understand the tunneling behaviour of an electrified interface.

The presence of low-lying electronic states in between the gap has been proposed as possible reason for the observed reduced tunneling barrier [143–146, 149, 158, 159]. The effect of such states is to *shorten* the effective tip/sample separation which leads to an apparent reduced barrier.

Such intermediate states arise from the structure of the solvent at the interface and have been identified with *solvated electrons* [144], *resonant water dipoles* [143, 145, 146, 149] and *resonating transient cavities* [158, 159].

Among the dynamic contributions of the solvent which influence the tunneling process, conformational fluctuations and two-dimensional diffusion of adsorbed species have been proposed [42].

The nature of the metals which constitute the junction is also relevant. It has been reported that for the same experimental conditions the tunneling barrier depends on the tip material [140–142].

6.2 Some particular aspect of the in-situ STM

But also the electrochemical behaviour of the substrate has to be considered. The electronic states directly involved in the tunneling process depend on the geometry of the surface, which, in turn, changes with the electrode potential. This is the case of the Au(111) electrode which has a reconstructed surface at potentials negative of the *pzc*. At the *pzc* the adsorption of anions from the solution assists the lift of the reconstruction and at more positive potentials the adsorbed anions undergo a order/disorder phase transition.

The tunneling barriers are determined with the experimental measure of the current as function of the distance between tip and substrate. This method is called *Distance Tunneling Spectroscopy* (DTS). This is an intrusive method which can disturb the measurements because the physical presence of the tip disrupts the local structure of the interface.

Furthermore, the electric field between tip and surface can have an aligning effect on the dipoles of the double-layer and it shouldn't be forgotten that at such a short distance the double-layers of tip and substrate come in contact. This gives rise to the so called *cross-talk* double-layer effect (see section 5.1) where the contact through the electrolyte with the reference electrode is lost and the local electrochemical conditions are not anymore well defined.

At present it is very difficult to estimate quantitatively all these tip-induced effects, hence more conveniently we speak about an *Effective Barrier Height* (EBH) which is experimentally accessible from the measured current curves through the relation:

$$\phi_{eff} = A \left(\frac{\partial \ln I_T}{\partial s} \right)^2 \qquad (6.2)$$

where $A = 0.952$ when the distances are measured in angstroms [153, 154, 160].

The experimental measurements of the EBH, its variations with potential, distance and substrate are used in the next sections in order to get detailed informations about the structure of the interface.

However, there is one more issue that needs to be discussed: what does the statement *tip/sample separation* really mean?

6.2.1 The problem of the absolute gap width

If not a direct measurement of the tunnel gap, which is at the moment not possible, at least an estimation of the real distance between tip and substrate is required for a meaningful discussion of the experimental data given by the tunneling spectroscopy.

CHAPTER 6. Local Analysis of the Electrified Interface

A strategy often used in the STM experiments is to bring the tip in contact with the surface and set the point where the current shifts from the tunneling to the ohmic regime as the real zero.

This procedure is not always possible because the electronics of usual commercial devices do not sustain the high currents required. In this very frequent case a realistic *estimation* of the gap is attempted.

A first approximation is to treat the tip/sample contact as a *Quantum Point Contact* (QPC), namely an ideal one-dimensional conduction channel with an opening much smaller than the Fermi wavelength, under the hypothesis of a ballistic transport and in absence of reflection [161].

To a QPC a quantum conductance $G = 2e^2/h$ is associated [162, 163], or alternatively a quantum resistance $1/G = 12.9 K\Omega$, a quantity depending only on fundamental constants and not on the nature of the materials.

The tip/substrate contact is far from behaving like an ideal QPC. Due to the wavelength of their electrons, metals are not able to form such kind of contacts: scattering of electrons (not ballistic transmission) and reflection cannot be avoided in a STM junction (these effects are reviewed in [164, 165]).

Furthermore, the tip has not an ideal single atom apex, and this yields a broader opening of the transmission channel, and impurities, defects and inhomogeneities (the tip is actually an alloy) contribute to a further scattering of the traveling electron.

Aware of these problems, Lang introduced an *effective contact resistance* $R = A \cdot h/e^2$, where the constant A is > 1 and depends on the material [166]. In spite of the very simple model he used, where the tip/substrate contact is treated in terms of a single atom shared by both sides of the gap, he was able to theoretically reproduce experimental data by extrapolation of the currents to an effective contact resistance of $R_{eff} = 35 K\Omega (A = 2.7)$ [153, 166].

Once the tip and substrate are in contact, the current doesn't change anymore with the distance and the tunneling barrier drops to zero. An extrapolation of the experimental barrier to zero could be an alternative method for the estimation of the absolute gap width, but the variation of the barriers with the distance doesn't allow for a simple and unambiguous procedure.

In the following, a direct fitting of experimental data with theoretical calculations is presented as a possible approach to the estimation of the real tip/sample distance.

6.3 Investigation of the Au(111)/H_2SO_4 interface

In Chapter 1 the crucial importance for electrochemists of a deep knowledge of the structure of the electrified interface has been discussed. In this section, an application of the distance tunneling spectroscopy for the study of the Au(111)/H_2SO_4 interface is presented.

In the DTS experiments, the distance dependence of the tunneling current for the Au(111) surface in H_2SO_4 has been measured. In the figure are shown

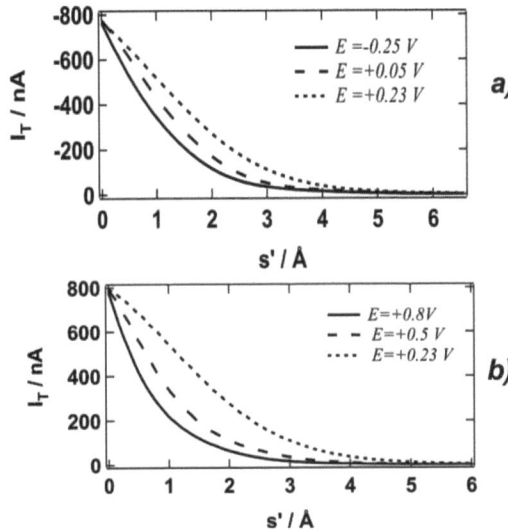

Figure 6.1: *Negative a) and positive b) tunnel currents I_T as function of tip distance s' for Au(111) in 0.1M H_2SO_4 at different electrode potential. $s' = 0$ is the closest approach experimentally achieved between tip and sample.*

the curves for potentials negative (fig. 6.1a) and positive (fig. 6.1b) of the pzc, as indicated in the picture. The tip potential was kept at +0.23 V/SCE, except for those cases where the sample potential itself was +0.23 V, that for the present system is just the pzc[113].

As a consequence, the various $I_T - s'$ curves were recorded for different tunnel voltages U_T. However, we considered constant electrochemical conditions at the tip surface to be of utmost importance. The origin $s' = 0$ in

CHAPTER 6.Local Analysis of the Electrified Interface

fig. 6.1 simply refers to an $I_T = 800nA$ for the bias indicated, the maximum value of the current experimentally accessible. The different electrode po-

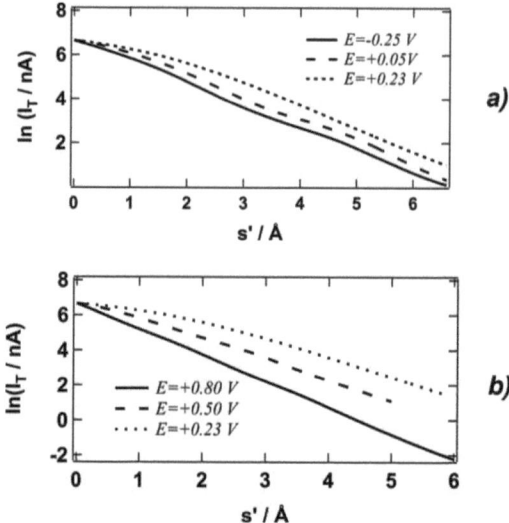

Figure 6.2: $ln(I_T)$ as function of tip distance s'. (From the experimental values in fig. 6.1).

tential values represent quite different situations for the double-layer which strongly influences the tunneling process occurring at the interface. Indeed, plotting $ln(I_T)$ vs. s' (fig. 6.2), the slope changes with the potential, and the non-exponential decay of the currents with the distance is evident. This behaviour of the currents gives rise to features in the calculated EBH which contain detailed information on the structure of the double-layer.

The structure of the double-layer at potentials positive of the pzc

At +0.8 V the sulfate ions are adsorbed on the Au(111) unreconstructed surface and form an ordered adlayer with a $(\sqrt{3} \times \sqrt{7})R19.1°$-structure [169–172]. In fig. 6.3 the relative EBH, directly calculated from the data shown in fig. 6.1 with Eq. 6.2, is shown. The origin $s' = 0$ corresponds to the distance from the surface where the currents is 800nA for a tunneling bias of 525 mV. Some preliminary consideration on the position of the tip must be done.

6.3 Investigation of the Au(111)/H_2SO_4 interface

In fig. 6.4 is evident that at 800nA the STM doesn't resolve the sulfate structure anymore but the atomic resolution of the underlying Au(111) surface can be recognized, although not clearly. At 800nA the tip goes through the adlayer and with its movement removes the adsorbed anions. At 2nA the sulfate structure is again visible, proving that in this case the tip is above the adlayer. It can be concluded that the origin $s' = 0$ for the current vs. distance curves of fig 6.2 corresponds to a position in between the sulfate adlayer. In other words, the tunneling current for +800 mV/SCE is measured *through* and immediately above the sulfate adlayer.

Although the periodicity of the sulfate adlayer is well studied, there is still not clear conclusion on the nature of the co-adsorbates, which have also been observed with the *in situ* STM (fig. 6.5).

An atomistic description of this structure based on DFT calculation has been provided by Dr. Timo Jacob and his co-worker Sudha Venkatachalam of the Institute of Electrochemistry at the University of Ulm.

In these DFT calculations the co-adsorption of water and hydronium in different configurations and structures have been considered. As one might expect, for the pH value used in the experiments, the calculations with co-adsorption of a single hydronium per unit cell showed the best agreement with the STM image and with the distance tunneling data shown in fig. 6.3.

Therefore, under these conditions the co-adsorption of bisulfate as well as water molecules can be ruled out. A top view of this system is shown in fig. 6.5a together with the corresponding experimental STM image fig. 6.5b, while a side view of a single unit cell together with the DFT analysis is shown in fig. 6.6.

Regarding the geometry, the lower three oxygen atoms of sulfate which

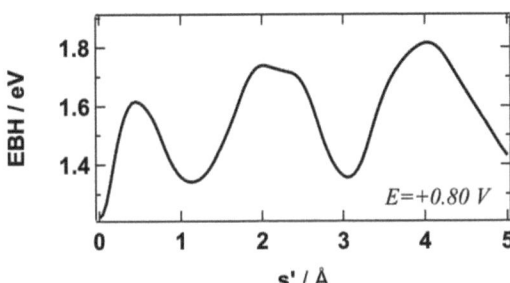

Figure 6.3: *Effective Barrier Height (EBH) vs. distance for Au(111) in 0.1M H_2SO_4. E_{tip}=+0.23 V/SCE*

CHAPTER 6.Local Analysis of the Electrified Interface

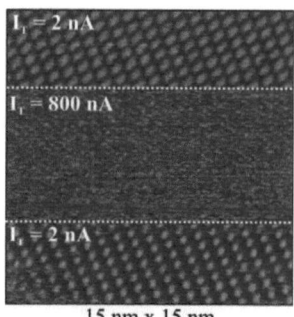

Figure 6.4: *STM image for Au(111) in 0.1 H_2SO_4 at different tunneling currents. E=+0.8 V/SCE, E_{tip}=+0.23V/SCE.*

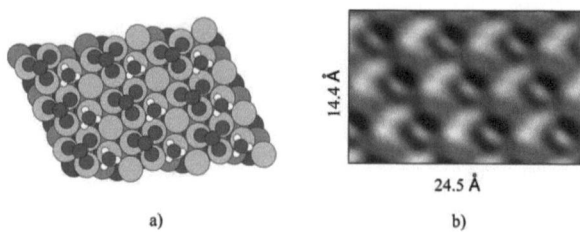

Figure 6.5: *a) Geometry optimized structure of ($\sqrt{3} \times \sqrt{7}$) sulfate being co-adsorbed with hydronium and b) the corresponding STM image at E=+0.8 V/SCE in H_2SO_4 0.1M*

6.3 Investigation of the $Au(111)/H_2SO_4$ interface

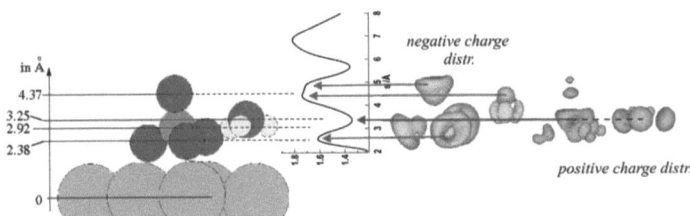

Figure 6.6: *Comparison of the DFT-calculated structure with the STM-measurements. Left: side view of a single ($\sqrt{3} \times \sqrt{7}$) unitcell including the vertical distances of the different atoms from the the first surface layer; Middle: experimental EBH with the lowest peak set in correspondence of the lower oxygens of sulfate; Right: Distribution of the negative and positive charges in the unitcell (Courtesy of T. Jacob and S. Venkatachalam).*

bind to the surface, are in average 2.38Å above the surface forming each a single covalent bond to the corresponding Au atom below (fig. 6.5a).

In addition, each of these O-atoms bind to the central S-atom with a $d(S-O) \approx 1.55$Å. While two of these O-atoms show an equivalent behaviour, the oxygen, which is also hydrogen-bonded ($d(O-H) = 1.74$Å) to two adjacent hydroniums, forms somewhat weaker S-O- and Au-O-bonds, which leads to a slight increase of both bond lengths.

While sulfate mainly orients with respect to the underlying substrate, the position of hydronium is determined by its ability to form hydrogen-bonds in a position which pushes its oxygen above the plane of the the O-sulfate (see fig. 6.6).

The STM-measured EBH is thought to originate from the charge distribution in the tunneling gap, and hence reflects the potential energy profile of the interface portion under investigation: an accumulation of negative charge hampers the electron transfer and results in a maximum in the EBH.

Based on this considerations, in order to relate the distance tunneling spectroscopy measurements to an absolute vertical position with respect to the surface plane, the first peak in the EBH in fig. 6.3 has been set in correspondence of the plane formed by the three sulfate oxygens which bind the surface for which the DFT-calculations showed an accumulation at the distance to the surface of these oxygens (fig. 6.6). The EBH with the corrected distance origin together with the relative STM image of the surface under

CHAPTER 6. Local Analysis of the Electrified Interface

investigation is reported in fig. 6.7.

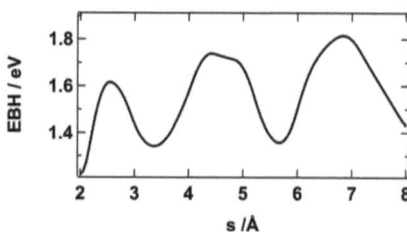

Figure 6.7: *STM image (left) and EBH vs. distance (right) for Au(111) in H_2SO_4 0.1M. E=+800mV/SCE. E_{tip}=+230mV/SCE*

On this distance scale the first minimum is 3.2 − 3.3Å above the surface plane, which interestingly coincides with the vertical position of the the oxygen of the hydronium.

This clearly shows that correlating the position of the maxima of the EBH with the oxygen in the system is not sufficient. However, there is a correlation with the charge density distribution, which a the distance of the first minimum shows a pronounced accumulation of positive charge (*electron depletion*) around the sulfur atom and the hydrogens of the coadsorbed hydronium.

By further withdrawing the STM tip from the surface, two, not completely resolved peaks close to each other at 4.3 and 4.8Å are found (fig. 6.7.

These two peaks find their direct correspondence in the calculated charge density distribution, which identifies the peak at 4.3Å as negative charge accumulated slightly above the oxygen atom of hydronium (occupying a *p*-orbital and the second peak as negative charge being located at the topmost O-atom of sulfate.

The DFT calculations were performed only for the specifically adsorbed sulfate and that's why the third peak of fig. 6.3 doesn't have a relative theoretical description as the first two, appearing at a distance from the surface which exclude specific adsorption.

Indeed, the minimum at around 5.5Å and the following maximum at around 6.9Å most probably reflect the next water layer, hydrating the adsorbates on the electrode and the charge distribution in the diffuse layer respectively.

An important goal has been achieved: the procedure above allows to give an explanation of the origin of the EBH a the electrified interface and at the same time the absolute width of the tunneling gap between tip and sur-

6.3 Investigation of the Au(111)/H_2SO_4 interface

face was estimated. The measurements and the complementary theoretical calculations of the charge distribution on the sulfate structure adsorbed on Au(111) can now be used as reference for the other potentials in order to achieve a detailed picture of the structure of the interface. Furthermore, the *real* zero of the curves shown above corresponds to an empirical contact resistance of $R_{cont} = 35K\Omega$, which is surprisingly the same value proposed by Lang in his theoretical model of the STM gap (see section 6.2.1). If this has a physical reason or it is just a coincidence is not possible to decide on the basis of the results presented here. However, these results are used in the following for the estimation of the distance between tip and surface for the others electrode potentials.

At +0.5 V/SCE the Au(111) surface is positively charged and unreconstructed, and gold island originated from the lift of the reconstruction are imaged with the STM (fig. 6.8 left). The electrochemical studies for this system indicate that at this potential the sulfate anions are specifically adsorbed although they are not immobilized in an ordered structure and rather diffuse on the Au surface. This is also the reason why is not possible to see the sulfate with the STM. The peaks shown in the EBH (fig. 6.8 right)

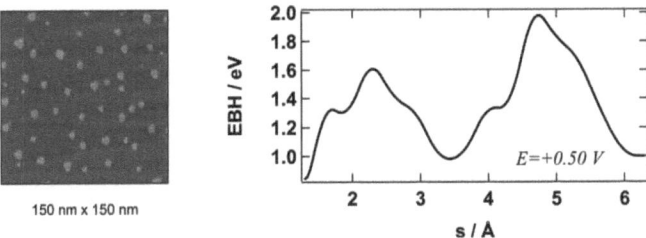

Figure 6.8: *STM image (left) and relative EBH vs distance (right) for Au (111) in H_2SO_4 0.1M. E=+0.500 V/SCE. E_{tip}=+230mV/SCE*

are almost in the same position of the first two peaks for the case of sulfate forming an ordered structure. The distance to the surface for the first peak can be assigned to species populating the IHP and this proves once again that sulfate is specifically adsorbed. The separation between the peaks is slightly larger than that shown in fig. 6.7. The small humps in the profile of the EBH are due to the stepped movement of the tip.

However, it must be noted that what is shown in fig. 6.8 is a time averaged value of the EBH, because the time scale of the measurements with the STM is much bigger than that of the 2-dimensional diffusion rate of sulfate on the surface. On the other hand, the presence of water a the surface and

CHAPTER 6.Local Analysis of the Electrified Interface

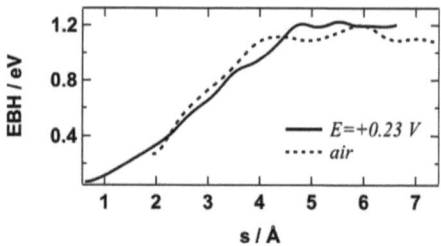

Figure 6.9: *Effective Barrier Height vs. distance for Au(111) in 0.1 H_2SO_4 at the pzc. E_{tip}=380mV/SCE. Also shown (dotted curve) is the EBH in air for freshly prepared Au(111). $U_t = 0.525V$.*

a different hydration shell around the anions can also influence the charge distribution through the gap.

At +0.23 V/SCE, the surface excess charge is practically zero (*pzc*). At the *pzc* the *oscillatory* behaviour shown for charged surfaces is completely absent, and the EBH increases almost linearly with the distance leveling off at about 1.2 eV for s>3-4 Å, the overall shape resembling the situation in UHV [155], although in the latter case the plateau reaches higher values (fig. 6.9).

Experiments performed with the same flame annealed Au(111) electrode in air yield almost identical barrier height (dotted curve in fig. 6.9). Because typical EBH values for gold in UHV range between 3 and 4 eV, it must be the water layer on the Au(111) electrode, which seems unavoidable for measurements performed in air, which brings the EBH down to the *in-situ* value of the uncharged surface.

It is well known that the liquid side of the interface at the *pzc* presents an *ordered* water structure (see section 1.2), the results shown in fig. 6.9 demonstrate that is the net charge distribution through the gap which gives rise to the features of the EBH, rather than the structure of the interface itself.

The results discussed above are shown together in 6.10 which can be seen as a snapshot of the structure of the Au(111)/H_2SO_4 interface at different positive electrode potentials. In the figure is also reported the EBH measured for the sulfate ordered structure at a tunneling bias which allows to the tip to get closer to the surface. It can be seen that the first two peaks are reproduced and the drop of the EBH in proximity of the point contact is clearly visible.

6.3 Investigation of the Au(111)/H_2SO_4 interface

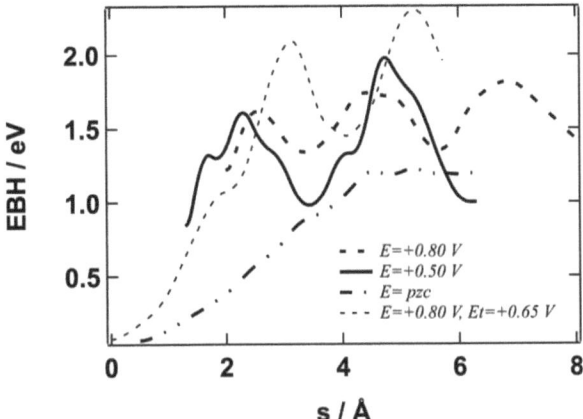

Figure 6.10: *EBH vs. distance for Au(111) in 0.1M H_2SO_4 at different electrode potentials.*

Double layer structure for potentials negative of the pzc

At -0.25 V/SCE the Au(111) electrode is negatively charged with cations at the Outer Helmholtz Plane and the surface is reconstructed (fig. 6.11 left)[113]. In the right side of the figure, the EBH for this potential is shown,

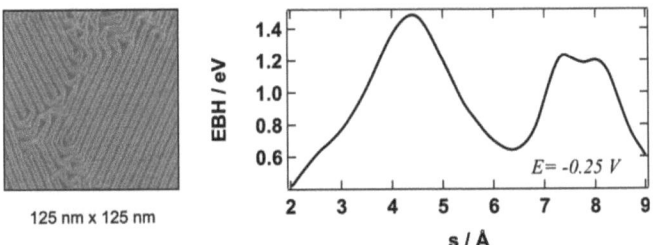

Figure 6.11: *STM image (left) and EBH (right) for Au(111) in 0.1M H_2SO_4. E=-0.25 V/SCE, E_{tip}=+0.23V/SCE*

with the absolute distance scale calculated by extrapolation to the contact resistance of $35K\Omega$.

It is worthy to note that the same estimation of the tip/sample distance

CHAPTER 6. Local Analysis of the Electrified Interface

was obtained with the general expression for the tunneling current Eq. 2.18 with the origin fixed at the same position as for the case of the sulfate adlayer and by substituting an average value of the EBH shown in fig. 6.11: the two approaches led to a difference in the absolute estimation of the gap of a fraction of angstrom. The result is shown in left part of fig. 6.11.

In order to obtain a similarly detailed picture of the interface at $E < pzc$ and to understand the features in the corresponding tunneling barrier, DFT calculation were performed for the adlayer forming on the negative surface charges.

In section 1.2 the formation of a well ordered water structure at the water/electrode interface had been discussed. For the present experimental conditions, the DFT calculations predict that for Au(111) the configuration in which the perpendicular water molecules are oriented with one hydrogen toward the surface is more stable than the H-up orientation. This behavior is in agreement with results for Pt(111), where the H-down configuration is slightly more stable [19].

Since the ordered region of the electrolyte includes more than a single water-bilayer, in the theoretical model a second bilayer has been included resulting in an ice-like (hexagonal) structure adsorbed on Au(111). This configuration further stabilizes the H-up orientation because allows for more additional hydrogen bonds.

As consequence of the contact between water-bilayer and electrode surface and the fact that the model-system contains two water layers only, the hydrogen bonds within the water layers (1.98 Å) and those connecting both (1.92Å) are elongated compared to the hydrogen bonds in hexagonal bulk-ice (exp.: 1.78Å [167]).

The adsorption of hydronium, that at this potential can't be neglected, is simulated by adding to the water molecules of the second bilayer an extra hydrogen neutral atom which after the calculation results to be charged positively.

In the following the theoretical calculation of the charge distribution arising from the structure described is compared with the experimental measure of the EBH. But in this case the estimation of the absolute tunneling width is already available.

The comparison of the EBH profile with the theoretical charge distribution is shown in fig. 6.12 where, as indicated by the arrows, the maxima of the EBH are clearly located in correspondence of the positive charge accumulation, and the minima originate from negative charge, yielding to a description of the origin of the EBH mirroring the case for the ordered sulfate structure on Au(111) at $E > E_{pzc}$.

For the STM-tip, which at $E < E_{pzc}$ is positively charged compared to the

6.3 Investigation of the Au(111)/H_2SO_4 interface

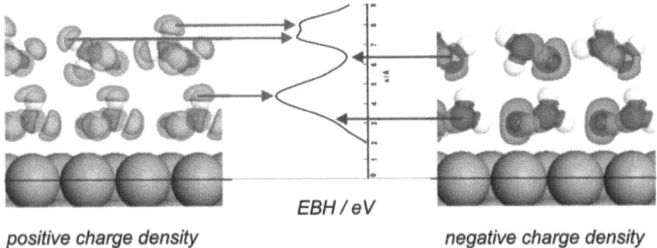

positive charge density   EBH / eV   negative charge density

Figure 6.12: *Comparison of the DFT-calculated charge distribution with the STM-measurements. Left: Side view of the two water-bilayers together with the positive charge density distribution. Right: Side view of the two bilayers together with the negative charge distribution. The correspondence between the charge densities and the extrema of the experimental EBH are indicate with arrows. (Courtesy of T. Jacob and S. Venkatachalam).*

electrode surface any nearby positive charge density (or electron deficiency) means an additional potential barrier, leading to a maximum in the EBH. Similarly, culmination of negative charge in the double-layer should appear as a minimum.

The first feature in the experimental curve at around 3.0Å off the surface plane comes from the competition between the positive charge located at the hydrogens and the negative charge located at the oxygens of the horizontally oriented water molecules of the first water bilayer.

The first maximum in the EBH-curve at around 4.3Å is due to the positive charge at the hydrogens of the perpendicular-oriented waters, which are facing away from the surface (H-up).

While from the calculations these H-atoms are 4.24Å away from the surface, the cloud of positive charge is significant between 4.2 and 4.6Å. Since at this distance from the surface there are no oxygen atoms which could lower the barrier, the first peak in the EBH is quite pronounced.

The following minimum around 6.2Å coincides with the strong accumulation of negative charge from the oxygen atoms of the second water-bilayer, oppressing any influence from positive charge at similar surface distance. This includes oxygen from the horizontally as well as perpendicularly oriented water molecules of the second bilayer.

CHAPTER 6. Local Analysis of the Electrified Interface

Quite interesting is the following peak that apparently shows two maxima at 7.4 and 7.9Å.

The outer feature at 7.9Å comes from H atoms at water molecules which are not influenced by the presence of hydronium. These water molecules of the second bilayer are oriented such that one of the H-atoms is facing away from the surface. The inner maximum of this peak at 7.4Å finds its corresponding positive charge located at the outer hydrogens of water molecules, which is due to hydronium, are arranged such that the inner hydrogens are facing toward the surface (H-down).

Therefore, the calculations show that although none of the maxima comes from a positive charge density located at the hydronium molecules directly, the double-feature structure of the outermost peak in the EBH-curve between 7.4 and 7.9Å is a direct consequence of their presence.

This finding was verified by calculations for the analogous system but without hydronium (two water-bilayers adsorbed on Au(111)) which showed no accumulation of positive charge density at 7.4Å above the surface and which is not able to fully describe the EBH-curve.

The presence of hydronium has been also reported by Lilach and co-workers for Pt(111) surface in contact with a rather thick layer of ice (>150 ML) [168]. Since in that work the electrode-near area of the electrolyte arranges in an ice-like structure too, there might be a relevance of the results discussed above even to non-electrochemical interfaces. At E=+0.05 V/SCE,

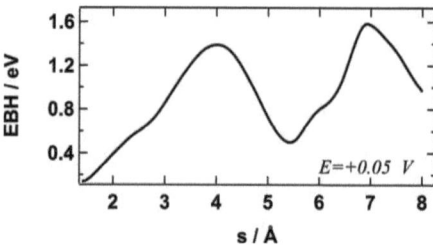

Figure 6.13: *EBH vs. distance for Au(111) in 0.1M H_2SO_4. E=+0.05 V/SCE, E_{tip}=+0.23V/SCE*

the Au(111) is still reconstructed and negatively charged. The profile of the relative EBH shown in fig. 6.13 reproduces that for E=-0.25 V/SCE. This experimental finding suggests that the water structure discussed for the more negative potential is retained also for this slightly less charged surface.

The comparison of the EBH measured at potentials negative of the *pzc* is reported in fig. 6.14 where the difference between the positions of the peaks

6.3 Investigation of the Au(111)/H_2SO_4 interface

is below the experimental error. The results presented here demonstrate how

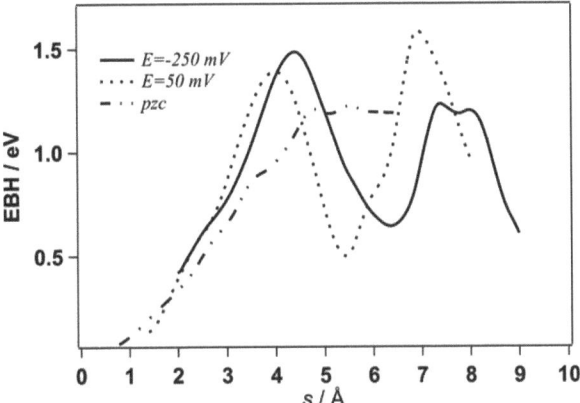

Figure 6.14: *Comparison of the EBH vs distance for Au(111) in 0.1M H_2SO_4 at potentials negative of the pzc.*

powerful the application of the distance tunneling spectroscopy can be for a detailed study of the interfaces, and not only those under potential control.

Since in the following sections the method and the results for Au(111) are used as reference for the analysis of other system, an outlook of the complete work is shown in fig. 6.15 together with a cyclic voltammetric curve. Also important, the results presented here allow for a better understanding of the tunneling process through an interface because the EBH profile discussed above can be substituted in Eq. 2.14 for the determination of the *faradaic* current flowing through the interface. And since the tunneling term directly affects the pre-exponential factor of the expression for the ET at the interface, it can be concluded that on the basis of the result of fig. 6.15, the exchange current i_0 is expected to change with the potential as consequence of a different structure of the interface which gives rise to different potential barrier through the gap (see section 2.2).

It must also be said that these variations maybe don't have any experimental relevance because the EBH present very close *average* values and the eventual correction could be negligible. On the other hand, an eventual variation of i_0 related to the change of the properties of the tunneling gap is an important information for the description of the processes of charge transfer at the electrochemical interfaces.

CHAPTER 6. Local Analysis of the Electrified Interface

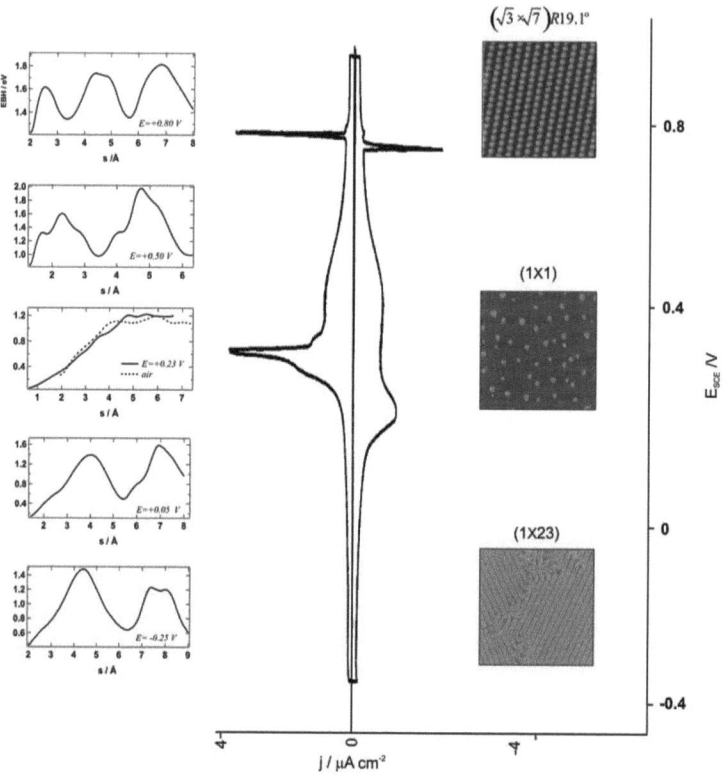

Figure 6.15: *CV curve (Middle) and EBH vs distance at different potentials for Au(111) in 0.1M H_2SO_4.*

6.4 An electrochemical nano-switch

The behaviour of the tunneling current at the Au(111)/H_2SO_4 interface has been also studied as function of the tunneling bias for fixed tip/substrate distance.

This method, which is called Voltage Tunneling Spectroscopy, or simply Tunneling Spectroscopy (TS) has been used in UHV studies in order to get detailed informations about the electronic structure of the substrate [173, 174].

In the present case, the *in-situ* STM setup allows to change independently the potential of both tip and surface and a tunneling bias is established just by choosing two different potentials for tip and substrate.

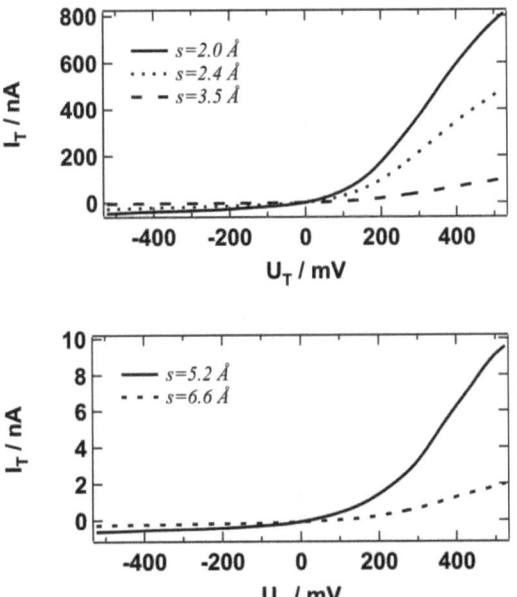

Figure 6.16: *Tunneling current vs. bias for Au(111) in H_2SO_4 0.1M. E_{tip} = +0.23V/SCE. (The distances s are determined from the DTS curves)*

It is evident that in this case the variation of the tunneling bias is achieved by changing the potential of only one side of the junction. In this way not

CHAPTER 6. Local Analysis of the Electrified Interface

only the electronic structure, but also the electrochemistry of the system determines obtained with this method that could be laso called *Potential Tunneling Spectroscopy*.

In the following, the results for the Au(111)/H_2SO_4 system obtained by changing the electrode potential and keeping fixed the tip potential at different fixed tip/substrate distances are reported.

Conceptually, the tunneling current as function of the bias could be obtained from the DTS measurements presented in the previous section just by reporting the values that the current assumes for different bias (*i.e.* different substrate potentials) at a chosen distance.

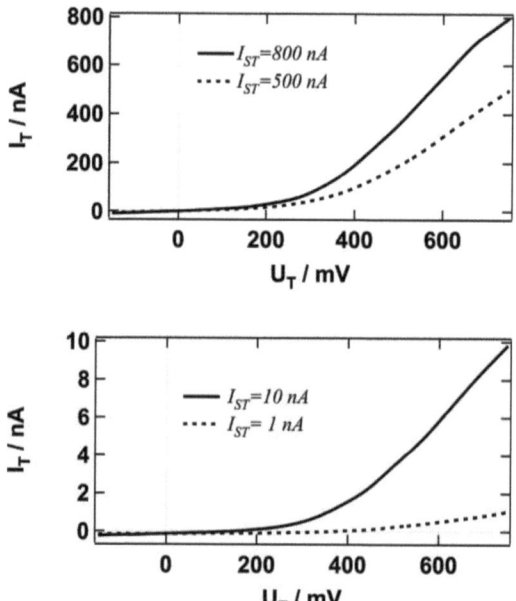

Figure 6.17: *Tunneling currents vs. bias for Au(111) in H_2SO_4 0.1M. E_{tip} = +0.05V/SCE. I_{ST} measured at E=+800mV/SCE.*

At a distance s=2Å from the surface held at +0.8 V/SCE, in fig. 6.1 the current is ≈ 800 nA for a tunneling bias of U_T=525 mV. When the electrode potential is set to -0.25 V/SCE, for the same distance, the current is ≈ -800 nA (U_T=-525 mV). Actually, a quite symmetrical behaviour is expected on

6.4 An electrochemical nano-switch

the basis of the DTS curves.

The TS measurements reveals a surprisingly different behaviour: the currents show a marked asymmetry when they are recorded versus the tunnel bias(Fig. 6.16). This behaviour must be somehow related to the *changing potential* since the DTS measurements made for fixed potentials do not reveal the phenomenon.

From what is shown in fig. 6.16, one could doubt that the asymmetry is due to the inversion of the bias, namely to the inversion of tunnel *from* occupied states at negative potentials to tunnel *into* empty states of the substrate at positive bias.

Asymmetrical behavior with respect to the zero bias point explained in terms of asymmetrical electric field through the gap have been already reported in the literature [149, 155, 173]. Although in the present case the difference between the currents on the right and left side is really high, the measurements were repeated for different tip potentials (different zero bias points) in order to make sure whether the inversion of the tunneling direction does really matter.

In fig. 6.17 the measurement for a tip potential of +50mV/SCE are shown. The asymmetry is reproduced but for a bias value where the electrode potential assumes the same value where the asymmetry in fig. 6.16 arises, independently on the inversion of the bias ($U_T = 0$ mV) point.

It becomes clear that the sign of the bias, namely the direction of the tunneling current don't play any role. This is once more demonstrated in fig. 6.18 where this time the zero bias has been shifted in the positive direction.

Very surprisingly, there is a region of higher bias (on the left) where the current is lower than for lower bias (between -400 and 0 mV). Moreover, in the region immediately around the zero bias, the current is symmetrical.

For all the examples shown, the asymmetry is reproduced for all the tip/substrate distances investigated. I remind that the distances are indicated as starting current set points (I_{ST}) at a fixed bias: the higher the I_{ST}, the closer the tip.

As already mentioned, in the actual experimental setup, the tunneling bias is changed by changing the Au(111) potential and keeping fixed the tip potential. It seems that the asymmetry arises from some modification of the sample related to variation of the electrochemical potential.

It is well known that up the *pzc* sulfate is specifically adsorbed at the Au(111) surface which undergoes a disorder/order phase transition up to +800 mV/SCE. The presence of an adsorbate modifies the electronic structure of the system and hence the pre-exponential factor of the STM-current which contains the density of states. But in the present case, the order of magnitude of the asymmetry of the currents is much too high to be explained

CHAPTER 6. Local Analysis of the Electrified Interface

in terms of Local Density of States (LDOS).

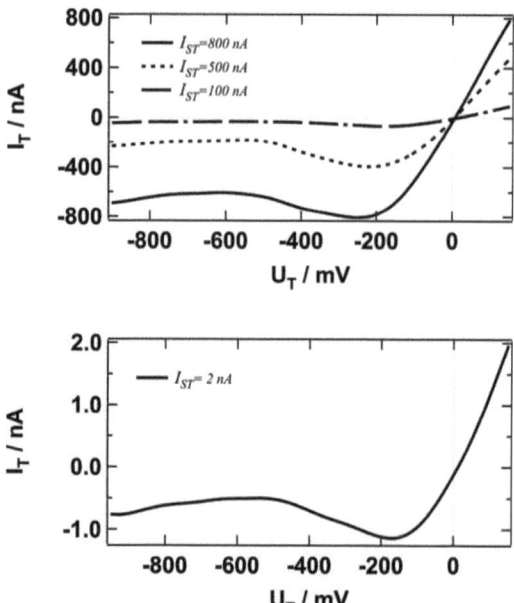

Figure 6.18: *Tunneling current vs. bias for Au(111) in H_2SO_4 0.1M. $E_{tip} = +0.65V/SCE$. I_{ST} measured at $E=+800mV/SCE$.*

Another possible explanation is that the adsorption of the sulfate brings molecules into the tunneling gap, resulting in a shortening of the tunneling gap at the *pzc*. In fig. 6.4 has been demonstrared that the tip can penetrate into the sulfate layer and can tunnel directly into the empty states of the gold surface. Hence, the adsorption of sulfate does not reduce the effective distance between tip and surface.

In order to have a reference system, the measurements have been repeated for a monolayer of Pd on Au in sulfuric acid, a system which presents the sulfate adsorption but doesn't reconstruct. The comparison with the results obtained for the Au(111) surface are shown in fig. 6.19.

The difference is eye catching: the lift of reconstruction of the gold surface must play a role in the strong asymmetry of the tunneling currents at different bias. It must be noted that for the Pd layer the desorption of sulfate at very

6.4 An electrochemical nano-switch

negative potentials (bias negative of -400 mV) doesn't affect the current. The possible effect of the reconstruction of the gold surface was also checked

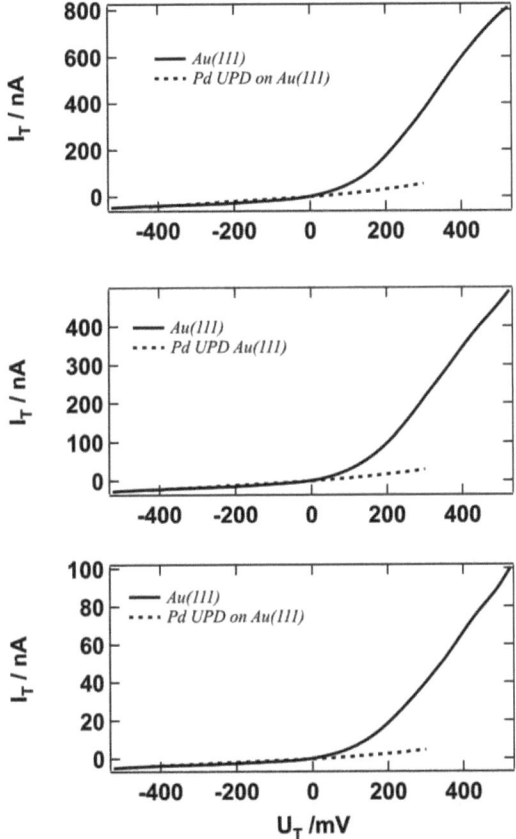

Figure 6.19: *Currents vs tunneling bias for Pd UPD on Au(111) (dotted) and bare Au(111) at different tip/sample distances. Electrolyte: H_2SO_4 0.1M. Et=+0.23 V/SCE. The measurements for the Pd UPD have been stopped before the onset of oxidation and dissolution.*

by measuring the tunneling currents for the same sample in air, namely by excluding the potential control which gives rise to the modifications of the

CHAPTER 6.Local Analysis of the Electrified Interface

surface. The results for two different tip/substrate distances together with the *in-situ* current are shown in fig. 6.20.

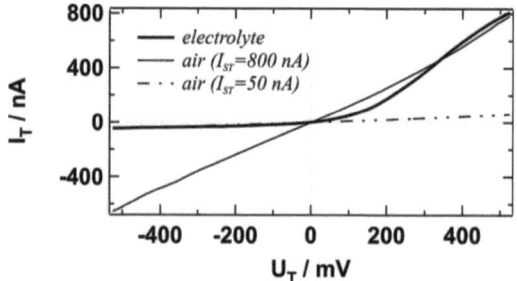

Figure 6.20: *Tunneling currents vs. bias for Au(111) in air a two different tip/sample separations (dotted and thin line) compared with the in situ measurements in H_2SO_4 0.1M.*

It can be seen that the two branches of the *in-situ* measurements can be fit with two curves measured in air a two different tip/sample distances. It seems that the asymmetry originates really from a variation of the distance between tip and surface as effect of the electrochemical potential.

From the DTS measurements (Fig. 6.1) it can be calculated that the current decays from a value of 800 nA (left side of fig. 6.20, electrolyte curve) to a value of 50nA (right side) in a distance of about 2.3Å, which is the height of a gold monoatomic step.

Let us try to give an explanation to these experimental findings. The reconstructed Au(111) surface is characterized by a slightly reduced lattice constant and presents a compressed structure with respect to the underlying bulk layers. This disposition of the atoms at the surface gives rise to the characteristic lines imaged with the STM (see the STM image on the left side of fig. 6.11). In correspondence of the *pzc* this structure is not anymore stable and the surface relaxes to the bulk lattice constant. The exceeding atoms are ejected from the first layer and form islands on the surface.

The process is very fast and if the tip is just above a forming island, the width of the gap results reduced by exactly a monolayer. A schematic representation of the effect of the lift of the reconstruction on the geometry of the tunneling gap is shown in fig. 6.21. This can give just a partial explanation of the experimental results and only in the case the tip is *accidentally* above a forming island. Furthermore, the TS currents don't show any hysteresis with the direction of the bias scan. That means that the destruction of the

6.4 An electrochemical nano-switch

islands should be a quick process, which is not.

Indeed, the reconstruction of surfaces involves the displacement of a large amount of atoms which requires the overcome of kinetic barriers. The thermal reconstruction is visible with the STM only after annealing the sample, and the so called potential-induced reconstruction at negative potentials needs several minutes in order to re-embed the atoms from the islands and give rise to the characteristic lines [113].

But, in order to redistribute the atoms of the surface as reaction to a different charge, the space where they will be located must be created. This can be a fast process occurring also on top of the islands, a mechanism which can be thought as a *breathing* of the atoms at the surface in response to the applied potential. This could yield a statistical different extension of the STM junction with a difference corresponding again to 2Å, namely an Au monoatomic step. This mechanism is sketched in fig. 6.22.

On the other hand, some modifications of the tunnel gap induced by the tip can not be excluded. These possible explanations need further investigations also on the theoretical side. However, the process behind the strong asymmetry on the tunneling currents is quick and reversible and if it is not correct to speak of a rectifying effect, we can refer to the Au(111)/H_2SO_4 interface as to an *electrochemical nano-switch*.

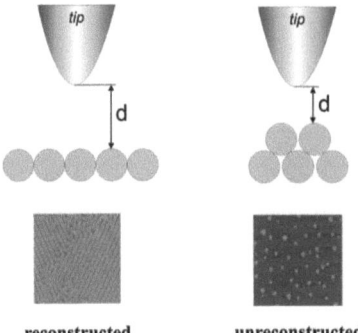

Figure 6.21: *Schematic representation of the shortening of the tunneling gap caused by the lift of the reconstruction of Au(111). The corresponding STM images are shown.*

CHAPTER 6.Local Analysis of the Electrified Interface

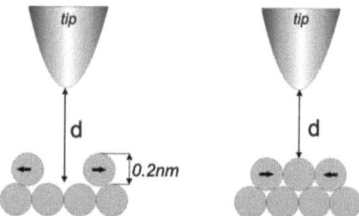

Figure 6.22: *Schematic representation of the* breathing *mechanism which yields to a variation of the width of the STM junction.*

6.4.1 Electronic properties of the interface

The TS measurements discussed above allow for some insight into the electronic properties of surfaces and their transformations with the applied potential.

The TS curves present features which are partially obscured by the exponential decay of the tunneling current. Most of this dependence can be removed by direct computation of the so called *normalized tunneling conductance* $(dI/dU_T) \cdot (U_T/I)$ which is a quantity related somehow to the LDOS (see section 2.3). Anyway, the real meaning of this parameter is still discussed in the literature [31, 173–176].

The magnified images of the respectively negative and positive bias portions of fig. 6.20 are reported in fig. 6.23. The deviations of the *in-situ* curves from the measurements performed in air are evident. In order to better describe the beahviour of these currents, the $(dI/d_T) \cdot (U_T/I)$ for Au(111) in sulfuric acid has been calculated from the experimental results and is shown in fig. 6.24.

A peak centered at $U_T = +200mV$ is clearly visible. It arises from the shortening of the tunneling gap caused by the lift of the reconstruction already discussed. A slight increase of the tunneling conductance is also seen at the negative extreme. The DTS measurements demonstrate that the water bilayers together with the adsorption at the OHP of hydronium ions also is retained also at potentials very negative of the *pzc*. The peak at -525mV is not due to a change in the structure of the electrolyte. By looking at the electrode potentials reported in the upper part of fig. 6.24 corresponding to the bias indicated in the bottom, the feature can be due to tunneling from an *occupied* state generated from the onset of the hydrogen adsorption and eventually evolution. This behaviour will be very evident in the tunneling spectroscopy experiments performed on pseudomorphic layers of Pd on

6.4 An electrochemical nano-switch

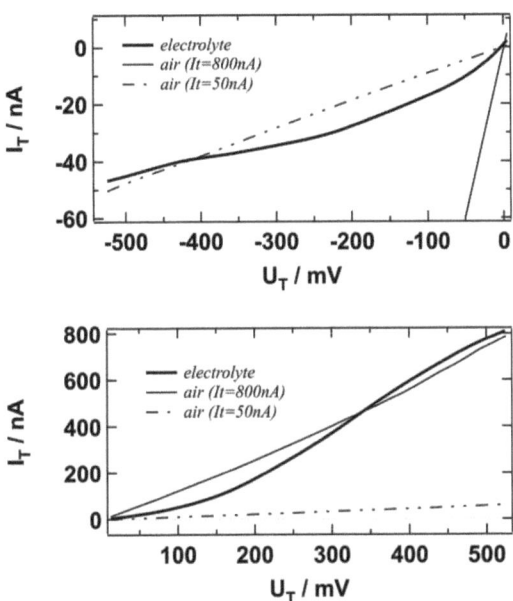

Figure 6.23: *Tunneling current vs. bias for Au(111) in H_2SO_4 0.1M. Magnifications of fig. 6.20.*

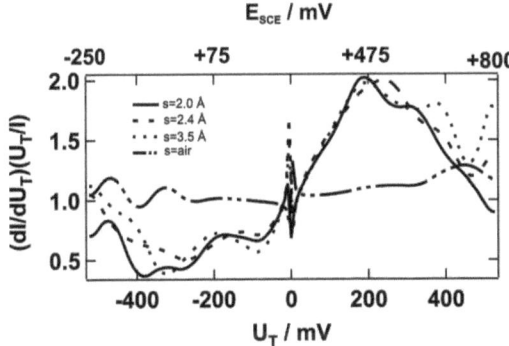

Figure 6.24: *Normalized Tunneling Conductance for Au(111) in H_2SO_4 0.1M. (Calculated form the results of fig. 6.16).*

CHAPTER 6. Local Analysis of the Electrified Interface

Au(111) discussed in the next section.

Anyway, the features of the tunneling conductance are reproduced for different tip potentials, as shown in fig. 6.25 where the shift in the y-axis is the effect of a shifted Fermi level of the tip at different potentials.

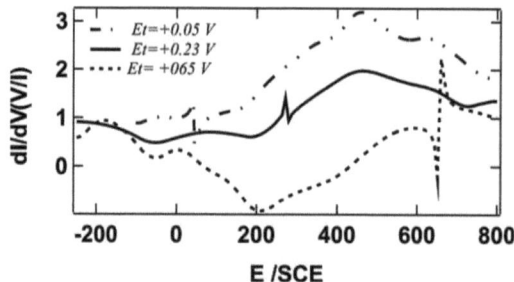

Figure 6.25: *Normalized Tunneling Conductance for Au(111) in H_2SO_4 0.1M at different tip potentials E_t.*

In order to describe the effects of both electrolyte and potential on the tunneling process, the $(dI/dU_T) \cdot (U_T/I)$ has been calculated also for Au(111) in air and is shown in fig. 6.24.

The measurements in air show a constant value of $(dI/dU_T) \cdot (U_T/I)$, an experimental finding which points to the electrochemical nature of the features appearing in the *in-situ* experiments.

In the right portion of fig. 6.24, for bias between +400mV and +525mV, the tunneling conductance decreases and a clear minimum appears as the tip/substrate distance increases.

In order to check whether this behaviour is an artifact, the tunneling spectroscopy measurements were performed by excluding the bias (potential) values where the asymmetry originates and for a starting current of $I_{ST} = 2nA$ (s=6.6 Å) which correspond to a farther distance of the tip from the surface. The corresponding tunneling conductance is shown by the solid curve in fig. 6.26.

The minimum is clearly reproduced. It can be also noticed that the profile of the solid curve before the minimum is very close to that of the tunneling conductance in air (solid thin curve) and this confirms that the peak at +200mV arises from geometrical transformation of the STM junction and not from the electronic configuration.

The depth of the minimum at +460mV decreases as the tip approaches the surface and disappears for tunneling current of 800nA (fig. 6.26).

6.4 An electrochemical nano-switch

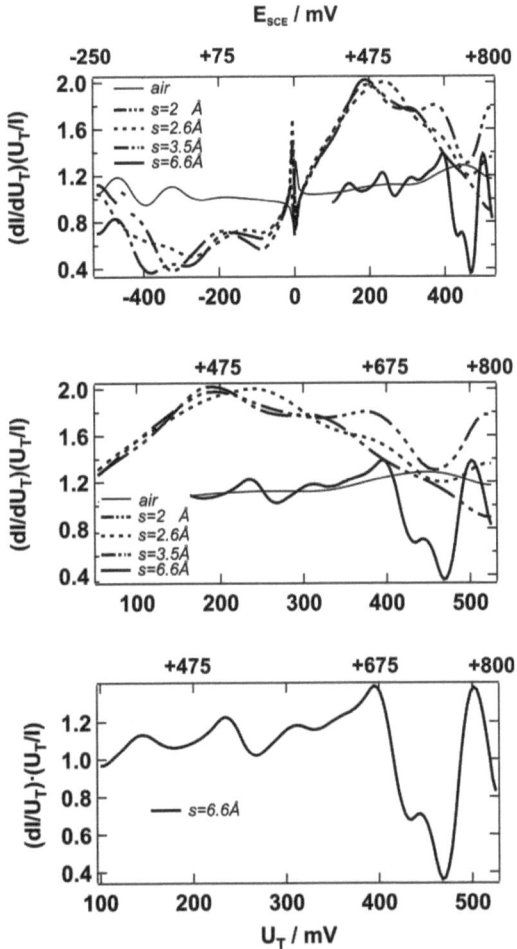

Figure 6.26: *Normalized tunneling conductance for Au(111) in H_2SO_4 0.1M at different tip/sample separations. Middle: detail of the positive bias region. Bottom: detail of the curve for s=6.6Å recorded for the only positive bias scan excluding the asymmetry point. E_{tip}=+230mV/SCE*

CHAPTER 6.Local Analysis of the Electrified Interface

As already shown in fig. 6.4, for a tunneling current of 800nA the sulfate adlayer is not imaged anymore with the STM and the underlying Au surface is visible although not very clear. In this configuration the electron tunnels from an occupied state of the tip directly into an empty state of the gold surface.

As the tip is retracted, there are a series of intermediate situations till the sulfate adlayer is imaged for currents lower of 10nA. In this case and just as consequence of the tip/surface separation, the electron tunnels from an occupied state of the tip into an empty state of the adsorbate. This situation could explain the distance dependence of the tunneling conductance.

More in general, the origin of the minimum itself can be due to the disorder/order transition of the sulfate which implies a reorganization of the electrons among HOMO and LUMO of both substrate and adsorbate. In other words, the feature at +460mV in the tunneling conductance originate from a particular electronic configuration of the substrate.

Although this kind of measurements could give important and detailed information about the general behaviour of the electrified interface, at the moment nothing more can be said without a theoretical modeling of the process.

What is really important in the curves shown in fig. 6.26 is that with the *in-situ* tunneling spectroscopy the features in the profile of $(dI/dU_T) \cdot (U_T/I)$ originate from the *transformation* of the electronic structure of the sample in response to an applied electrochemical potential, and this makes the difference with the usual measurements in UHV conditions, where the aim is to get information on a static, frozen electronic structure.

6.5 Spectroscopic investigation of Pd thin films

6.5.1 Structure of the interface with the DTS

Thin films of Pd on Au(111) have attracted the attention of electrochemists for their structural and catalytical properties. In this section both the distance and voltage tunneling spectroscopy methods are used for the investigation of the interface at a Pd-UPD layers on Au(111) in a 0.1M H_2SO_4 solution. A general description of the electrochemical behaviour of Pd thin film has already been provided in section 5.3.1.

In fig. 6.27 the deposition of two successive Pd layers is shown. The well known $\sqrt{3} \times \sqrt{7}$ structure of the sulfate adlayer covering the Pd-UPD surface is recognizable in fig. 6.27c.

After the deposition of two Pd monolayers, the cell has been removed

6.5 Spectroscopic investigation of Pd thin films

from the STM in order to exchange the electrolyte with a Pd free H_2SO_4 0.1M solution.

The cell has been then placed again into the microscope and the potential control reactivated. Fig. 6.28a was recorded immediately after the re-assembling. Although some white spot could be due to the deposition of Pd clusters in an uncontrolled way during the transfer of the cell, the well ordered sulfate structure demonstrates that the Pd film retains its pseudomorphic structure.

For spectroscopic measurements, mechanical vibrations arising from the handling of the system must be avoided in order to don't affect the extremely sensitive measurements. That's why is not possible to measure immediately after the exchange of the electrolyte. Moreover, the present experiments are performed after the oxygen has been removed from the cell, and this procedure needs several minutes too. The figures 6.28b and 6.28c reveal that after the waiting time for damping down the mechanical vibrations and for removing the oxygen, the sulfate adlayer doesn't show anymore the $\sqrt{3} \times \sqrt{7}$ structure. In fig. 6.28b some ordered domain is still visible but after a shorter time the sulfate adlayer appears everywhere like in fig. 6.28c. In order to better visualize the sulfate adlayer, the first derivative of the STM image reported in fig. 6.28c is shown in fig. 6.29.

The reason of such transformation of the sulfate structure could be that the underlying Pd film is not anymore in registry with the gold surface, in other words, is not anymore pseudomorphic. However, the analysis of the structure of the Pd layers on Au goes beyond the scope of the present work for which the important information is that sulfate is present at the surface and the measurements reported in the following refer to the structure imaged in fig. 6.28c and fig. 6.29.

Figure 6.27: *STM images of Pd deposition on Au(111) from H_2SO_4 0.1M + $PdSO_4$ 0.1mM. a) Growth of the first Pd UPD, b) Growth of a 2nd Pd layer, c) detail of the sulfate adlayer on the Pd thin film.*

CHAPTER 6. Local Analysis of the Electrified Interface

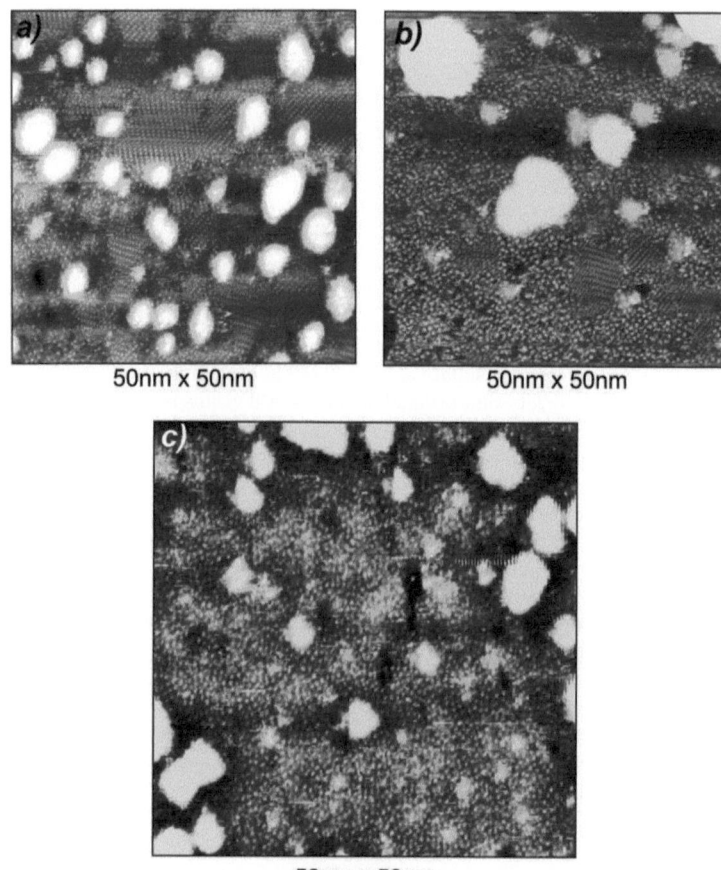

Figure 6.28: *STM image of the sulfate adlayer on Pd-UPD after the exchange of the electrolyte. The three images have been recorded a different times. E=+0.1 V/SCE, electrolyte: H_2SO_4 0.1M.*

6.5 Spectroscopic investigation of Pd thin films

The range of stability of the sulfate adlayer on Pd allows to measure the tunneling currents also at tip potentials positive of the sample potential.

In fig. 6.30 the tunneling current variations with the tip/sample distance (note that origin of the s' axis refers to the maximum value of the currents under the actual experimental conditions) are shown for two different tip potentials, one referring to tunneling *from the tip* to the sample (positive current, dotted line) and the other for tunneling *from the sample* to the tip (negative current, full line). The different values of the bias give rise to a small difference between the two curves in the figure, which in turn reveals no asymmetry in the tunneling process from and to the sample.

The absolute width of the tip/surface separation were determined by extrapolating the experimental currents to the operative contact resistance of $35 K\Omega$.

Hence, the effective tunneling barrier was calculated directly from the experimental data with Eq. 6.2. The results are shown in fig. 6.31.

For the case of sulfate adsorbed on an Au(111), it has been already explained that the features in the EBH originate from the charge distribution through the gap, and then the curves shown in fig. 6.31 demonstrate once more that the specific adsorption of sulfate anions on Pd thin films gives rise to a charge *layering* also in the case of a not well ordered structure of the

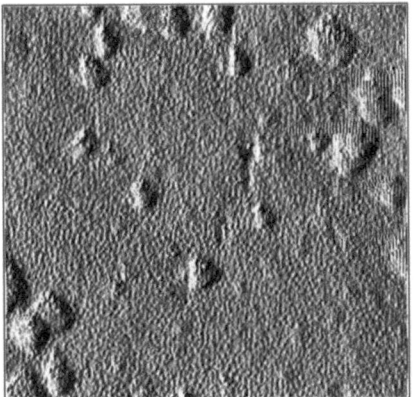

50nm x 50nm

Figure 6.29: *First derivative of the STM image shown in fig. 6.28c.*

CHAPTER 6. Local Analysis of the Electrified Interface

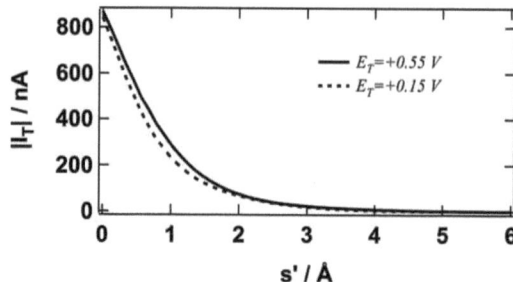

Figure 6.30: *Tunneling currents vs. distance for Pd UPD on Au(111) for two different tip potentials. E=+0.40 V/SCE, electrolyte: H_2SO_4 0.1M.*

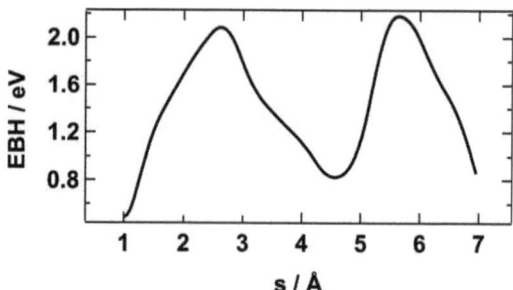

Figure 6.31: *EBH vs tip/sample separation for Pd-UPD on Au(111). E=+0.40 V/SCE, electrolyte: H_2SO_4 0.1M*

6.5 Spectroscopic investigation of Pd thin films

adlayer. The two peaks are very similar to those discussed for the sulfate adsorption on Au(111). Only the outer third peak appearing in the measurements on the $\sqrt{3} \times \sqrt{7}$ structure is missing (compare with fig. 6.7). In fact, this extra peak has been explained as arising from a negative charge layer due to an ordered structure of the electrolyte at the OHP just above the adlayer.

One can conclude that the unordered sulfate layer on the Pd layer shown in fig. 6.28c doesn't induce any structure on the surrounding electrolyte.

Quite surprisingly, the EBH of fig. 6.31 is very similar to that reported in fig. 6.8, and a direct comparison is made in fig. 6.32. It could be concluded

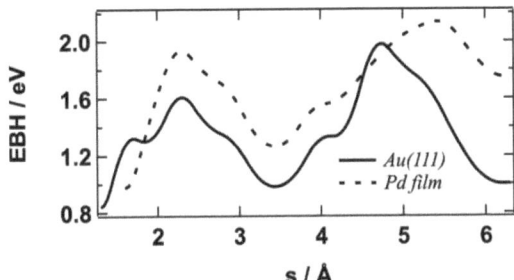

Figure 6.32: *EBH vs distance for Au(111) at E=+0.50 V/SCE and Pd-UPD at +0.40 V/SCE in H_2SO_4 0.1M*

that the time-averaged structure of the disordered sulfate on Au(111) at electrode potentials under +800mV/SCE is more similar to the sulfate adlayer on a Pd monolayer shown in fig. 6.28c than to the well ordered $\sqrt{3} \times \sqrt{7}$ shown in fig. 6.8.

A last consideration regards the slight difference in the height of the maxima. Indeed, the value of the maxima in the EBH calculated for the unordered structure on Pd films are higher than those for the disordered structure on Au(111) at +0.5 V/SCE and also higher than those for the $\sqrt{3} \times \sqrt{7}$ structure at +800mV/SCE (fig. 6.7).

The presence of hydronium coadsorbed with sulfate has been demonstrated to be necessary for the stabilization of the $\sqrt{3} \times \sqrt{7}$ structure. As can be seen in fig. 6.6, hydronium introduces additional charge in the adsorbate and it results in a lower charge separation: in fig. 6.6 it can be seen that the negative charge contribution overlaps with the positive charge coming from the sulfate.

These considerations together with the experimental STM evidences for

CHAPTER 6. Local Analysis of the Electrified Interface

sulfate on Pd UPD yield to the conclusion that the not ordered structure shown in fig. 6.28 could be due to the absence of coadsorbed hydronium.

6.5.2 Voltage Spectroscopy investigation

The behaviour of the tunneling current vs. tunneling bias for Pd thin films on Au(111) has been already shown in section 6.4.1. In the following a more detailed discussion is provided.

As for the case of Au(111), the *in-situ* Tunneling Spectroscopy measurements are performed by keeping the tip potential fixed at +0.23 V/SCE and scanning the electrode potential.

The results for different tip/samples separations are shown in fig. 6.33. I remind that the distances are reported as different starting currents (set-points I_{ST}). Since the starting points are set for the same bias, the higher the current, the closer the tip.

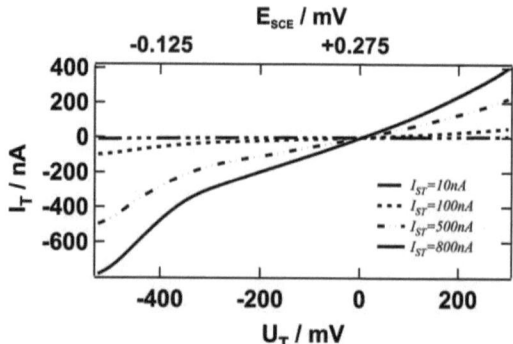

Figure 6.33: *Tunneling current vs. bias for different tip/substrate separations for Pd-UPD/Au(111) in H_2SO_4 0.1M*

The Pd thin films on Au(111) do not reconstruct, then the strong variation of the current decay at very negative bias (which means also very negative electrode potentials) can't be explained in terms of effective variation of the gap width.

In order to get more information on the system, the normalized conductance $dI/dU_T \cdot (U_T/I)$ for the curves shown in fig. 6.33 has been calcualted and is reported in fig. 6.34.

The broad feature appearing at negative bias can't be assigned to some electronic state: it extends for too many mV.

6.5 Spectroscopic investigation of Pd thin films

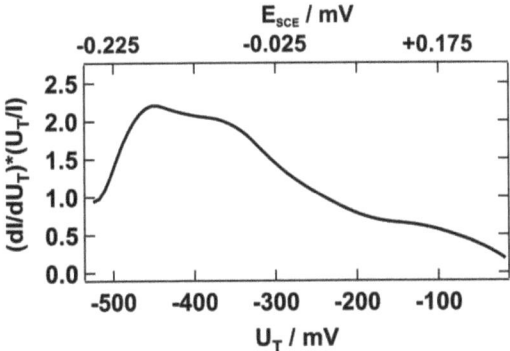

Figure 6.34: *Normalized Tunneling Conductance vs. bias for Pd-UPD(Au(111) in H_2SO_4 0.1 M. Etip=+0.23 V/SCE*

It is well known that on Pd pseudomorphic layers, the adsorption of hydrogen occurs contemporary to the desorption of sulfate between -0.05 V and -0.10 V/SCE [112, 123]. When more than 1ML are deposited on Au(111), the hydrogen is absorbed at further negative potentials till the hydrogen evolution starts around -0.30 V/SCE.

In fig. 6.33 a variation in the current slope is clearly seen around -0.10 V, resulting in a maximum in the normalized conductance in fig. 6.34. It can been concluded that the adsorption (and eventually the absorption) of hydrogen enhances the tunneling current and it can been explained in terms of an increase of the occupied states of the substrate arising from the adsorption process.

Nevertheless, a component of the currents due to the hydrogen evolution can't be completely excluded. At potentials around -250mV/SCE the hydrogen forming at the Pd layer diffuses toward the tip which is kept at a potential where hydrogen is reoxidized, resulting in an additional current detected at the tip, as sketched in fig. 6.35.

Although this effect gives rise to an anodic current at the tip, in our microscope a flux of electron from the solution to the tip is defined as negative current. Therefore, hydrogen evolution at the surface can bend the tunneling current recorded at the tip in the negative direction.

By summarizing the results so far presented, we can demonstrate that the spectroscopic methods are powerful tools for the investigation of electrochemical systems although a complementary theoretical work is necessary for a better understanding of the experimental findings. While the DTS is

CHAPTER 6.Local Analysis of the Electrified Interface

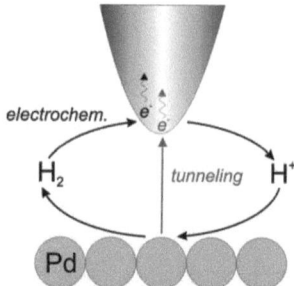

Figure 6.35: *Schematic representation of the electrochemical (black) and tunneling (red) components of the current flowing through an STM tip.*

able to provide detailed informations about the structure of the interface, the VTS, if not the electronic density of states is suitable for the local detection of processes occurring at the interface.

6.6 First investigations of Pd nano-clusters

The aims of this Ph.D. thesis were on one side the application of the STM for the local modification of surfaces based on the jump-to-contact phenomenon (Chapter 5), and the development of the *in-situ* tunneling spectroscopy for the local investigation of the electrified interface.

In this section the two roads can finally meet together and the different chapters composing this work can get their synthesis: all the work presented and discussed along the entire thesis aimed to write this short section. It's not by accident that this is also the last one.

In the previous sections the tunneling spectroscopic methods have been used for a detailed study of the structure of the electrochemical double-layer at different potentials. Although the reported results refer to the properties of the surface as an all, what has in practice been done is to probe the small region between tip and substrate: in other words, the surface has been locally probed!

The question is now whether the tunneling spectroscopy can be used for the study of the local properties of small heterogeneities on a large surface, otherwise not accessible with the classical methods of electrochemistry.

In order to explore the limits of such possibility, the method has been used for the investigation of a small Pd cluster on a gold surface.

6.6 First investigations of Pd nano-clusters

The cluster was generated in the same way as described in Chapter 5 where an STM-tip, previously loaded with Pd in different cell, was approached to the surface till the Jump-to-contact occurred. Working with a preloaded tip allows to measure in a Pd-free solution.

An image of the the cluster used for the experiments is shown in fig. 6.36 together with a side view. It must be reminded, that in the figure the

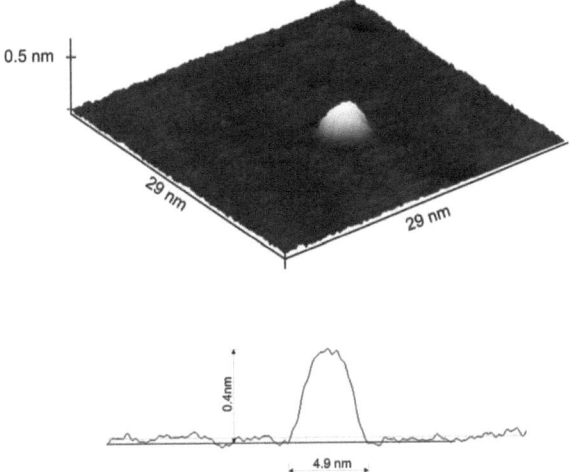

Figure 6.36: *STM image of a Pd cluster on Au(111) surface. The cluster was generated with the jump-to-contact method with a tip preloaded with Pd. E=-50 mV/SCE, electrolyte: H_2SO_4 0.1M*

apparent size of the cluster is shown (see chapter 5), and the real diameter at the basis of the cluster is thought to be smaller.

On the other hand, the resolution of the tunneling spectroscopy depends on the curvature radius of the tip, which can be larger than the dimensions of the cluster itself. This can give already an idea about the local resolution of the tunneling spectroscopies.

Indeed, the calculation of the effective tunnel barriers for the system shown in fig. 6.36 reveals features which are not reproducible (fig. 6.37). On the other hand, the EBH profiles on top of the cluster are not featureless as in the case of the *pzc* of a Au(111) surface. Minima and maxima can be observed.

CHAPTER 6. Local Analysis of the Electrified Interface

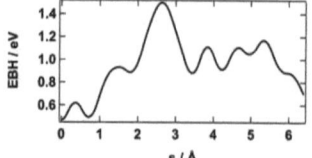
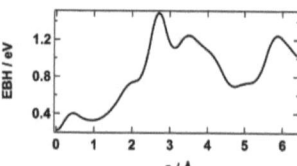

Figure 6.37: *EBH vs. distance from the point of highest tunneling current for a Pd cluster on Au(111). The two image refers to two different measurements. The features in the EBH profile are not reproducible. E= -150 mV/SCE, Et=+370 mV/SCE. Electrolyte: H_2SO_4 0.1M*

If it assumed that the curvature radius of the tip is bigger or of the same size of the cluster, then the tip can have a strong disturbing effect on the double-layer structure just above the cluster and on the disposition of species eventually adsopbed on the cluster.

The intrusive movement of the tip can move apart the molecules populating the IHP and OHP giving rise to the feature shown in fig. 6.37. When a flat surface is studied with the DTS the structure of the double-layer all around the investigated portion is retained also when the tip physically goes through an adsorbed layer, as in the case of sulfate.

In the case of the cluster, if sulfate is eventually adsorbed on top of it, the movement of the tip simply pushes away the molecules giving rises to the features shown in fig. 6.37.

It can be concluded that on the basis of the DTS measurements there are species adsorbed on top of the cluster, although is not possible to determine how molecules and their associated charged are distributed within the tunneling gap.

More promising seem to be the Voltage (or *Potential*) Tunneling Spectroscopy.

The measurements on Pd thin films show very clearly that the tunneling current is influenced by processes occurring at the surface.

The possibility to detect processes occurring on a nano-sized portion of the surface is demonstrated in fig. 6.38. The tip/sample distances are indicated as initial currents for the same bias of -200 mV: the higher the current, the closer tip.

It can be seen that the curves bend down for bias around than -200mV. The humps evident in all curves can be due to small vibrations of the systems, to which the STM is extremely sensitive due the special geometry of the junction.

As already seen for Pd thin films, the behaviour of the currents for this

6.6 First investigations of Pd nano-clusters

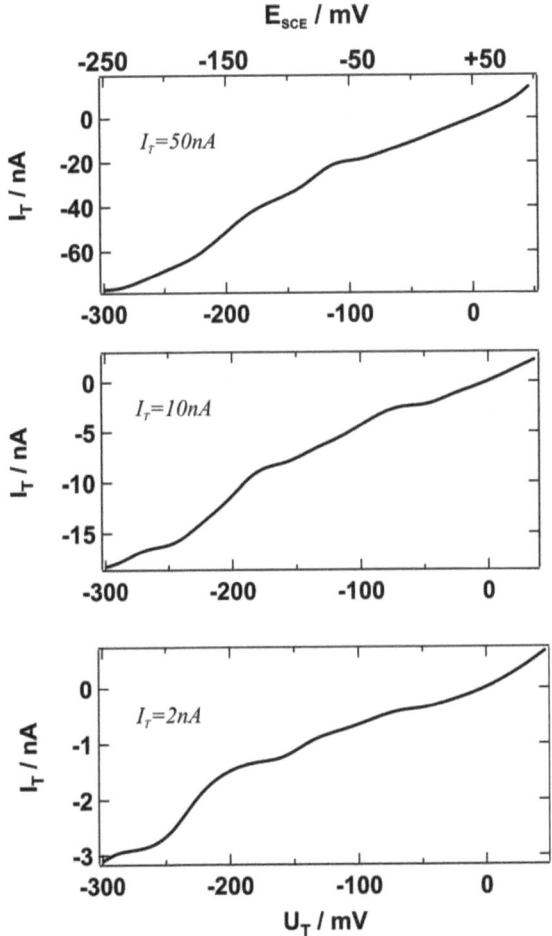

Figure 6.38: *Tunneling current vs. bias on top of a Pd cluster on Au(111) at different distances. Et=+50 mV/SCE. The distance are reported as current set points I_T at a bias=-200 mV (E=-150 mV/SCE)*

CHAPTER 6. Local Analysis of the Electrified Interface

negative range of potential (the electrode potentials corresponding to the applied bias are indicated on the upper axis in fig. 6.38) can be assigned to the hydrogen evolution at the cluster which is oxidized at the tip. This behaviour is reproduced for all the tip/sample separations.

Due the reduced dimensions of the system under investigation, detailed informations on the electronic structure of the cluster are extremely important. Such information can be obtained only by extending the bias window, but either the cluster or the electrolyte stability limit the range of the available potentials.

The measurements shown in fig. 6.38 clearly reveal an electrochemical activity of the Pd cluster. Although is not possible to get any quantitative information, the proof of hydrogen evolution at the clusters allows for meaningful further investigations on the electrocatalytical activity of such systems. An issue of crucial importance for both scientific and technological reasons.

Chapter 7

Summary

The topic of this thesis is the use of the *in-situ* STM beyond the imaging of electrode surfaces.

In Chapter 4 the study of a faceted surface of an Ir(210) single crystal has been reported.

Planar surfaces characterized by an high density of broken bonds are replaced by another one with a lower total surface energy and a larger extension due to the formation of facets with different orientations.

The driving force for such transformation is a strong surface energy anisotropy which is responsible for the evolution of the system (the surface) toward a thermodynamically more stable state in a way which is in close relation with the transitions between thermodynamic phases like, *e.g.* liquid and gas.

In the case of a faceted Ir(210) surface, the STM measurements have revealed the presence of pyramids showing two (311) and one (110) facets.

The study of extended portions of the surface reveals the presence of partially faceted domains which allows for some consideration about the faceting mechanism.

This kind of informations is crucial for a meaningful description of the electrochemical behavior of an electrode. Indeed, sharp peaks in the cyclic voltammograms can be assigned to the adsorption of anions which build an ordered structure only on the (110) nano-facet of the pyramids.

The STM has been also used for the modification of the surfaces with the controlled deposition of metal nano-clusters. The method is based on the so called Jump-to-contact between tip and substrate and is described in Chapter 5.

By applying a given electrode potential to the STM tip, metal can be deposited on it from the solution. The metal-loaded tip is made to approach the surface close enough, so that the Jump-to-contact occurs. This leads to

CHAPTER 7. Summary

the formation of a connective metal neck which is broken upon retracting of the tip. A small cluster is left on the surface and the tip can be reloaded again because of the ongoing metal deposition. The process can be repeated at will, also in the range of kHz.

The tip-generated clusters contain less that 100 atoms and show a surprisingly high stability against the anodic dissolution.

By optimizing this technique, large portions of the electrode surface were covered with nano-clusters making possible the analysis of these samples with other methods.

The influence of the presence of the nano-clusters on the metal deposition process were also studied. It has been observed that the clusters act like nucleation centers and retain a memory of the process upon dissolution of the deposits.

In Chapter 6 the results of a Tunneling Spectroscopy study of the electrified interface is presented.

In the Distance Tunneling Spectroscopy (DTS) the current is measured at different tip/sample distances while the surface is held at a constant potential. The Effective Tunneling Barrier can be calculated from the experimental results. With the help of theoretical calculations, the EBH has been directly related to the charge distribution in the STM gap yielding detailed informations about the structure of the interface at different potentials.

These measurements allowed also for an estimation of the absolute width of the STM gap.

Complementary to the DTS is the Voltage Tunneling Spectroscopy (VTS) where the current is measured as function of the electrode potential at a fixed distance. Although this method is used in UHV in order to get informations on the electronic structure of the substrate, its application at the liquid/solid interface allows for the study of the transformation of the electronic structure with the applied potential. Furthermore, the VTS has revealed processes occurring at the interface which was not possible to detect with the classical techniques used in electrochemistry. Indeed, a rectifying effect of the $Au(111)/H_2SO_4$ interface on the tunnel current has been discovered. Moreover, the overlap of tunnel and faradaic currents during VTS experiments performed on top of a tip-generated Pd cluster gave a first evidence of their electrocatalytical activity.

Chapter 8
Zusammenfassung

Thema der vorliegenden Arbeit ist die Verwendung der in-situ Rastertunnelmikroskopie (engl. scanning tunneling microscopy - STM) über die bloße Abbildung von Elektrodenoberflächen hinaus.

In Kapitel 4 wird die Untersuchung der facettierten Oberfläche eines Ir(210) Einkristalls dargestellt. Die relativ offene Struktur der planaren Ir(210)-Oberfläche ist dabei durch die Bildung von Facetten unterschiedlicher Orientierung in eine andere Struktur mit niedrigerer Oberflächenenergie umgewandelt.

Die treibende Kraft für eine derartige Umwandlung ist die hohe Anisotropie in der Oberflächenenergie, die dafür verantwortlich ist, dass sich das System (die Oberfläche) in Richtung eines thermodynamisch stabileren Zustandes entwickelt. Dieses Verhalten ähnelt typischen Phasenübergängen, wie z.B. zwischen Flüssigkeit und Gas. Im Falle einer facettierten Ir(210)-Oberfläche, haben STM-Messungen das Vorliegen von Pyramiden mit (311)- und (110)-Orientierung aufgezeigt. Die Untersuchung von größeren Oberflächenbereichen zeigt nur teilweise facettierte Domänen, die über den Mechanismus des Facettierungsprozesses Auskunft geben.

Diese Art von Informationen ist für eine aussagekräftige Beschreibung des elektrochemischen Verhaltens einer Elektrode äußerst wichtig. In der Tat können scharfe Strompeaks im Voltammogramm Adsorptionsprozessen auf den nano-Facetten der Pyramiden zugeschrieben werden.

Das Rastertunnelmikroskop (STM) wurde darüber hinaus für die Modifizierung von Oberflächen durch kontrollierte Abscheidung von Metall Nanoclustern eingesetzt. Diese Methode basiert wie in Kapitel 5 beschrieben auf dem so genannten *jump-to-contact* zwischen Spitze und Probe.

Durch Anlegen eines bestimmten Elektrodenpotentials an der STM Spitze kann Metall an dieser abgeschieden werden. Die mit Metall beladene Spitze wird dann nahe genug an die Oberfläche angenähert, so dass der *jump-to-*

CHAPTER 8. Zusammenfassung

contact ermöglicht wird. Dies führt zur Ausbildung eines so genannten *connective metal neck*, welches beim Zurückziehen der Spitze aufgebrochen wird. Auf der Oberfläche bleibt ein kleiner Cluster zurück, die Spitze kann aufgrund der anhaltenden Metallabscheidung neu beladen werden. Dieser Prozess kann nach Belieben bis in den KHz-Bereich wiederholt werden. Diese durch die STM-Spitze generierten Cluster bestehen aus weniger aal 100 Atomen und zeigen eine überraschend hohe Stabilität gegen anodische Auflösung.

In Kapitel 6 werden Ergebnisse zur Tunnel-Spektroskopie an der Grenzfläche Metall-Elektrolyt präsentiert. Bei der abstandsabhängigen Tunnelspektroskopie (DTS) wird bei konstantem Elektrodenpotential der Strom für verschiedene Abstände zwischen Spitze und Probe gemessen. Aus den experimentellen Daten kann die effektive Tunnelbarriere berechnet werden. Mithilfe von theoretischen Berechnungen wurde die effektive Tunnelbarriere direkt mit der Ladungsverteilung zwischen Spitze und Probe in Beziehung gesetzt. Dabei konnten detaillierte Informationen über die Struktur der Grenzfläche als Funktion des Elektrodenpotentials gewonnen werden.

DTS und potentialabhängige Tunnelspektroskopie (VTS), wobei der Strom bei festem Abstand als Funktion des Elektrodenpotentials gemessen wird, ergänzen sich gegenseitig. Obwohl diese Methode im Ultrahochvakuum (UHV) eingesetzt wird, um Informationen über die elektronische Struktur des Substarts zu erhalten, erlaubt die Anwendung für die Grenzfläche Metall-Elektrolyt die Untersuchung von Veränderungen der elektronischen Struktur mit dem angelegten Potential.

Darüber hinaus konnten mit VTS Grenzflächenprozesse aufgeklärt werden, die mit klassischen elektrochemischen Methoden nicht zugänglich sind. Tatsächlich wurde am System Au(111)/H2SO4 ein verstärkender Effekt auf den Tunnelstrom entdeckt. Zudem ergab der Überlapp von Tunnelstrom und Faraday Strom bei VTS Experimenten an mit der STM-Spitze generierten Pd Clustern erste Hinweise für deren elektrokatalytische Aktivität.

Bibliography

[1] S. Trasatti *J. Electroanal. Chem.* 123 (1981) 121

[2] J. Lipkowski and P. N. Ross editors *Structure of electrofied interfaces*; 1998, New York

[3] J. Maier and D. M. Kolb (editors), *Solid State Ionics* 94 (1997) 1

[4] R. Smoluchowski, *Phys. Rev,* 60 (1941) 661

[5] R. Guidelli and W. Schmickler, *Electrochim. Acta* 45 (2000) 2317

[6] J. R. Smith, *Phys. Rev.* 31 (1969) 522

[7] N. D. Lang and W. Kohn, *Phys. Rev. B* 1 (1970) 4555; 3 (1971) 1215

[8] J. P. Badiali, M. L. Rosinberg and J. Goodisman, *J. Electroanal. Chem.* 143 (1983) 73; 150 (1983) 25

[9] M. A. Henderson, *Surf. Sci. Rep.* 46 (2002) 1

[10] J. O'M Bockris and A. K. N. Reddy, *Modern electrochemistry*; 1970, New York

[11] P. A. Thiel and T. E. Madey, *Surf. Sci. Rep.* 7 (1987) 211

[12] D. Doering and T. E. Madey, *Surf. Sci.* 123 (1982) 305

[13] K.-I. Ataka, T. Yotsuyanagi and M. Osawa, *J. Phys. Chem* 100 (1996) 10664

[14] K.-I. Ataka and M. Osawa, *Langmuir* 14 (1998) 951

[15] M. F. Toney, J. N. Howard, J. Richer, G. L. Borges, J. G. Gordon, O. R. Melroy, D. G. Wiesler, D. Yee and L. B. Sorensen, *Nature* 368 (1994) 444

BIBLIOGRAPHY

[16] M. F. Toney, J. N. Howard, J. Richer, G. L. Borges, J. G. Gordon, O. R. Melroy, D. G. Wiesler, D. Yee and L. B. Sorensen, *Surf. Sci* 335 (1995) 326

[17] W. Schmickler, D. Henderson and O. R. Melroy, *Chem. Phys. Lett.* 216 (1993) 424

[18] E. Spohr in R. C. Alkire and D. M. Kolb (editors), *Advances in Electrochemical Science and Engineering Vol. 6*, Weinheim, 1999

[19] T. Jacob and W. A. Goddard III, *J. Am. Chem. Soc.* 126 (2004) 9360

[20] D. L. Price and J. W. Halley, *J. Chem. Phys.* 102 (1995)6603

[21] O. Stern, *Z. Elektrochem.* 30 (1924) 508

[22] D. C. Grahame, *Chem. Rev.* 41 (1947) 441

[23] R. Parsons in P. Delahay (ed.) *Advances in Electrochemistry and Electrochemical Engineering*, New York, 1961

[24] J. Bardeen, *J. Phys. Rev. Lett.* 6 (1961) 57

[25] R. A. Marcus, *J. Chem. Phys.* 49 (1965) 679

[26] H. Gerischer, *Z. Phys. Chem* 6 (1960) 223

[27] R. W. Gurney, *Proc. Roy. Soc. A (London)* 134 (1931) 137

[28] J.E. Andersen, A. Kornyshev, A. M. Kuznetov, L. L. Madsen, P. Moller and J. Ulstrup, *Electrochim. Acta* 42 (1997) 819

[29] G. Binnig and H. Rohrer, *Hel. Phys. Acta* 55 (1982) 726

[30] G. Binnig and H. Rohrer, *IBM Res. Develop.* 44 (2000) 279

[31] J. Tersoff and D. R. Hamann, *Phys. Rev. Lett.* 50 (1983) 1998

[32] J. Tersoff and D. R. Hamann, *Phys. Rev. B* 31 (1985) 805

[33] J. G. Simmons, *J. Appl. Phys.* 34 (1963) 1793

[34] J. Tersoff and N. D. Lang, *Phys. Lett. Rev.* 65 (1990) 1132

[35] D. Lawunmi and M. C. Payne, *J. Phys. Condensed Matter* 2 (1990) 3811

[36] C. J. Chen, *Phys. Re. Lett.* 65 (1990) 448

BIBLIOGRAPHY

[37] E. Tekman and S. Ciraci, *Phys. Rev. B* 42 (1990) 1860

[38] H. Ou-Yang, B. Källebring and R. A. Marcus, *J. Chem. Phys.* 98(1993) 7565

[39] H. Ou-Yang, R. A. Marcus and B. Källebring, *J. Chem. Phys.* 100 (1994) 7814

[40] C. L. Claypool, F. Faglioni, W. A. Goddard III, H. B. Gray, N. S. Lewis and R. A. Marcus, *J. Phys. Chem. B* 101 (1997) 5978

[41] F. Faglioni, C. L. Claypool, N. S. Lewis and W. A. Goddard III, *J. Phys. Chem. B* 101 (1997) 5996

[42] A. M. Kuznetov and J. Ulstrup, *Electrochim. Acta* 45 (2000) 2339

[43] C. H. Reinsch, *Numerische Mathematik* 10 (1967) 177

[44] M. Sander, R. Imbihl, J. V. Barth and G. Ertl, *Surf. Sci.* 271 (1992) 159

[45] K. Pehlos, T. E. Madey, J. B. Hannon and G. L. Kellogg, *Surf. Sci. Rev. Lett.* 5 (1999) 767

[46] I. Ermanoski, W. Swiech and T. E. Madey, *Surf. Sci.* 592 (2005) L299

[47] E. D. Williams and N. C. Bartelt, *Ultramicr.* 31 (1989) 36

[48] C. Herring, *Phys. Rev.* 82 (1951) 87

[49] J. G. Che, C. T. Chan, C. H. Kuo and T. C. Leung, *Phys. Rev. Lett.* 79 (1997) 4230

[50] I. Ermanoski, C. Kim. S. P. Kelty and T. E. Madey, *Surf. Sci.* 596 (2005) 89

[51] R. A. Campbell, J. Guan, T. E. Madey, *Catal. Lett.* 27 (1994) 273

[52] R. Barnes, I. M. Abdelrehim and T. E. Madey, *Top. Catal.* 14 (2001) 53

[53] W. Chen, I. Ermanoski and T. E. Madey, *J. Ame. Che. Soc.* 127 (2005) 5014

[54] I. Ermanoski, K. Pelhos, W. Chen, J. S. Quinton and T. E. Madey, *Surf. Sci.* 549 (2004) 1

[55] S. Motoo and N. Furuya, *J. Electroanal. Chem.* 167 (1984) 309

BIBLIOGRAPHY

[56] N. Furuya and S. Koide, *Surf. Sci.* 226 (1990) 221

[57] N. M. Markovic, R. R. Adzic, B. D. Cahan and E. B. Yeager, *J. Electroanal. Chem.* 377 (1994) 249

[58] T. Pajkossy, L. A. Kibler and D. M. Kolb, *J. Electroanal. Chem.* 600 (2007) 113

[59] M. Sanchez Cruz, M. J. Gonzalez Tejera and M. C. Villamanan, *Electrochim. Acta* 30 (1985) 1563

[60] D. M. Kolb, *Angew. Chem. Int. Ed.* 40 (2001) 1162

[61] D. M. Kolb, *Surf. Sci.* 500 (2002) 722

[62] D. M. Eigler and E. K. Schweizer, *Nature* 344 (1990) 524

[63] M. F. Crommie, C. P. Lutz and D. M. Eigler, *Science* 262 (1992) 218

[64] G. Meyer, L. Bertels, S. Zöphel, E. Heinze and K. H. Rieder, *Phys. Rev. Lett.* 78 (1997) 1512

[65] G. Meyer, L. Bartels, S. Zöphel and K. H. Rieder, *Appl. Phys. A* 68 (1999) 125

[66] M. T. Cuberes, R. R. Schlitter and J. K. Gimzewski, *Surf. Sci.* 371 (1997) L231

[67] F. M. Leibsle, *Surf. Sci* 514 (2002) 33

[68] C. Lebreton and Z. Wang, *Microelectronic Engineering* 30 (1996) 391

[69] D. M. Kolb, G. E. Engelmann and J. C. Ziegler, *Angew. Chem. Int. Ed.* 39 (2000) 1123

[70] D. M. Kolb, G. E. Engelmann and J. C. Ziegler, *Sol. Sta. Ion.* 131 (2000) 69

[71] D. M. Kolb and F. C. Simeone, *Electrochim. Acta* 50 (2005) 2989

[72] D. M. Kolb and F. C. Simeone, *Curr. Opin. Sol. State Mat. Science* 9 (2005) 91

[73] W. Li, J. A. Virtanen and R. M. Penner, *J. Phys. Chem.* 96 (1992) 6529

[74] R. M. Nyffenegger and R. M. Penner, *Chem. Rev.* 97 (1997) 1195

[75] M. Takai, H. Andoh, H. Miyazaki and T. Tsuruhara, *Microelectronic Engineering* 35 (1997) 353

[76] T. Homma, C. P. Wade and C. E. D. Chidsey, *J. Phys. Chem. B* 102 (1998) 7919

[77] T. Homma, N. Kubo and T. Osaka, *Electrochim. Acta* 48 (2003) 3115

[78] N. Kubo, T. Homma, Y. Hondo nad T. Osaka, *Electrochim. Acta* 51 (2005) 834

[79] J. R. LaGraff and A. A. Gewirth, *J. Phys. Chem.* 98 (1994) 11246

[80] M. Petri and D. M. Kolb, *Phys. Chem. Chem. Phys* 4 (2002) 1211

[81] Z.-X- Xie and D. M. Kolb, *J. Electroanal. Chem.* 481 (2000) 177

[82] Z.-X. Xie, X. W. Cai, J. Tang, Y. A. Chen and B. W. Mao, *Chem. Phys. Lett.* 322 (2000) 219

[83] W. Jan, Z. X. Cao, C. J. Zhou, J. Y. Kang and B. W. Mao, *Chem. Phys. Lett.* 373 (2003) 575

[84] R. Widmer and H. Siegenthaler, *Electrochem. Comm.* 7 (2005) 421

[85] V. Kirchner, X. H. Xia and R. Schuster, *Acc. Chem. Res.* 34 (2001) 371

[86] M. Koch, V. Kirchner and R. Schuster, *Electrochim. Acta* 48 (2003) 3213

[87] R. Schuster and G. Ertl in A. Wieckowski. E. Savinova, C. Vayenas (editors), *Catalysis and electrocatalysis at nanoparticle surfaces*, New York, 2003

[88] A. J. Bard and M. V. Mirkin (editors), *Scanning Electrochemical Microscopy*, New York, 2001

[89] J. Zhou, Y. Zu and A. J. Bard, *J. Electroanal. Chem.* 491 (2000) 22

[90] N. Baltes, L. Thouin, C. Amatore and J. Heinze, *Angew. Chem. Int. Ed.* 43 (2004) 1431

[91] J. Ufheil, C. Hess, K. Borgwarth and J. Heinze, *Phys. Chem. Chem. Phys.* 7 (2005) 3185

[92] R. T. Pötzschke, G. Staikov, W. J. Lorenz and W. Wiesbeck, *J. Electrochem. Soc.* 146 (1999) 141

BIBLIOGRAPHY

[93] W. Schindler, P. Hugelmann and J Kirschner, *J. Electrochem. Soc.* 148 (2001) C124

[94] W. Schindler, P. Hugelmann and F. X. Kärtner, *J. Electroanal. Chem.* 522 (2002) 49

[95] G. Binnig, C. F. Quate and Ch. Gerber, *Phys. Rev. Lett.* 56 (1986) 930

[96] J. M. Soler, A. M. Baro, N. Garcia and R. Rohrer, *Phys. Rev. Lett.* 57 (1986) 444

[97] S. Ciraci, A. Baratoff and I. P. Batra, *Phys. Rev. Lett. B* 42 (1990) 7618

[98] R. Wiesendanger, *Scanning Probe Microscopy and Spectroscopy*, Cambridge, 1994

[99] J. B. Pethica nad P. Sutton, *J. Vac. Sci. Technol. A* 6 (1998) 2490

[100] U. Landman, W. D. Luedtke, N. A. Burnham and R. J. Colton, *Science* 248 (1990) 454

[101] U. Landman and W. D. Luedtke, *J. Vac. Sci. Technol. B* 9 (1991) 414

[102] M. W. Ribarsky and U. Landman, *Phys. Rev. B* 38 (1998) 9524

[103] G. Gross, A. Schirmeisen, A. Stalder, P. Gütter, M. Tschudy and U. Dürig, *Phys. Rev. Lett.* 80 (1998) 4685

[104] M. Marsical, C. F. Narambuena, M. G. Del Popolo and E. P. M. Leiva, *Nanotechnology* 16 (2005) 974

[105] L. Kuipers and J. W. M. Frenken, *Phys. Rev. Lett.* 70 (1993) 3907

[106] L. Kuipers, M. S. Hoogeman and J. W. M. Frenken *Surf. Sci.* 340 (1995) 231

[107] Z. Gai, X. Li, B. Gao, R. G. Zhao, W. S. Yang and J. W. M. Frenken, *Phys. Rev. B* 58 (1998) 2185

[108] D. M. Kolb, R. Ullmann and T. Will, *Science* 275 (1997) 1097

[109] D. M. Kolb, R. Ullmann and J. C. Ziegler, *Electrochim. Acta* 43 (1998) 2751

[110] J. C. Ziegler, G. E. Engelmann and D. M. Kolb, *Z. Phys. Chem.* 208 (1999) 151

BIBLIOGRAPHY

[111] G. E. Engelmann, Ph.D. Thesis, University of Ulm, 1997

[112] J. Tang, M. Petri and D. M. Kolb, *Electrochim. Acta* 51 (2005) 125

[113] A. S. Dakkouri and D. M. Kolb in A. Wieckowski (editor) *Interfacial Electrochemistry*, New York, 1999

[114] M. A. Schneeweiss and D. M. Kolb, *Phys. Sta. Sol. A* 173 (1999) 51

[115] S. S. Sheiko, M. Möller, E. M. Reuvekamp and H. W. Zandbergen, *Ultramicroscopy* 53 (1994) 371

[116] P. Markiewicz and M. C. Goh, *J. Va. Sci. Technol. B* 13 (1995) 115

[117] C. Hess, *Hochauflösende Metalabscheidung mit dem elektrochemischen Rastermikroscop.*, Ph.D. thesis at University of Freiburg, 2004

[118] A. J. Bard and M. V. Mirkin (editors), *Scanning Electrochemical Microscopy*, New York, 2001

[119] P. Sung, F. O. Laforgue and M. V. Mirkin, *Phys. Chem. Chem. Phys.* 9 (2007) 802

[120] J. Heinze, *Angew. Chem. Int. Ed. Engl.* 32 (1993) 1268

[121] J. Clavilier, *J. Electroanal. Chem.* 107 (1980) 211

[122] J. Clavilier, R. Faure, G. Guinet and R. Durand, *J. Electroanal. Chem.* 107 (1980) 205

[123] M. Baldauf and M. M. Kolb, *Electrochim. Acta* 38 (1993) 2145

[124] M. Baldauf and D. M. Kolb, *J. Phys. Chem.* 100 (1996) 11375

[125] L. A. Kibler, A. M. El-Aziz and D. M. Kolb, *J. Mol. Catal. A* 199 (2003) 57

[126] A. M. El-Aziz and L. A. Kibler, *J. Electroanal. Chem.* 534 (2002) 107

[127] H. Naohara, S. Ye and K. Uosaki, *Coll. Surf. A* 154 (1999) 201

[128] L. A. Kibler, M. Kleinert, R. Randler and D. M. Kolb, *Surf. Sci.* 443 (1999) 19

[129] M. Takahashi, Y. Hayashi, J. Mizuki, K. Tamura, T. Kondo, H. Naohara and K. Uosaki, *Surf. Sci.* 461 (2000) 213

BIBLIOGRAPHY

[130] A. Ruban, B. Hammer, P. Stoltze, H. L. Skriver and J. K. Norskov, *J. Mol. Catal. A* 115 (1997) 421

[131] E. Christoffersen, P. Liu, A. Ruban, H. L. Skriver and J. K. Norskov, *J. Catal.* 199 (2001) 123

[132] A. Roudgar and A. Gross, *J. Electroanal. Chem.* 548 (2003) 121

[133] B. E. Koel, A. Sellidj and M. T. Paffet, *Phys. Rev. B* 46 (1992) 7846

[134] A. Roudgar and A. Gross, *Surf. Sci.* 559 (2004) L180

[135] C. G. Sanchez, E. P. M. Leiva and W. Schmickler, *Electrochem. Comm.* 5 (2003) 584

[136] M. G. Del Popolo, E. P. M. Leiva, H. Kleine, J. Meier, U. Stimming, M. Marsical and W. Schmickler, *Electrochim. Acta* 48 (2003) 1287

[137] W. J. Lorenz, G. Staikov, W. Schindler and W. Wiesbeck, *J. Electrochem. Soc.* 149 (2002) K-47

[138] C. Kittel, *Introduction to Solid State Physics*, New York, 1976

[139] K.-H. Meiwei-Broer (editor), *Metal Clusters at the Surface*, Berlin, 2000

[140] G. Nagy, *Electrochim. Acta* 40 (1995) 1417

[141] G.Nagy, *J. Electroanal. Chem.* 409 (1996) 19

[142] G. Nagy and G. Denuault, *J. Electroanal. Chem.* 437 (1997) 37

[143] J. Pan, T. W. Jing and S. M. Lindsay, *J. Phys. Chem.* 98 (1994) 4206

[144] J. K. Sass and J. K. Gimzewski, *J. Electroanal. Chem.* 308 (1991) 333

[145] S. M. Lindasy in H. Seigenthaler and A. A. Gewirth (editors) *Proc. NATO ASI on nanoscale Probes of the solid/liquid interface*, Netherlands, 1994

[146] A. Vaught, T. W. Jing and S. M. Lindsay, *Chem. Phys. Lett.* 236 (1995) 306

[147] M. Binggeli, D. Carnal, R. Nyffeneger, H. Siegenthaler, R. Christoph and H. Rohrer *J. Vac. Sci. Technol. B* 9 (1991) 1985

[148] R. Hiesigen, D. Eberhardt and D. Meissner, *Surf. Sci.* 597 (2005) 80

BIBLIOGRAPHY

[149] J. Halbritter, G. Repphun, S. Vinzelberg, G. Staikov and W. J. Lorenz, *Electrochim. Acta* 40 (1995) 1385

[150] M. Hugelmann and W. Schindler, *J. Electrochem. SOc.* 151 (2004) 1

[151] G. Nagy and T. Wandlowski, *Langmuir* 19 (2003) 10271

[152] Y. A. Hong, J. R. Hahn and H. Kang, *J. Chem. Phys.* 108 (1998) 4367, Errata Corrige *J. Chem. Phys.* 119 (2003) 9966

[153] J. K. Gimzewski and R. Möller, *Phys. Rev. B* 36 (1987) 1284

[154] N. D. Lang, *Phys. Rev. B* 37 (1988) 10395

[155] Y. Kuk nad P. J. Silverman *J. Vac. Sci. Technol. A* 8 (1990) 289

[156] W. Schmickler and D. Henderson, *J. Electroanal. Chem.* 290 (1990) 283

[157] W. Schmickelr, *Surf. Sci.* 335 (1995) 416

[158] U. Peskin, A. Edlund, I. Bar-On, M. Galperin and A. Nitzan, *J. Chem.Phys.* 111 (1999) 7558

[159] M. Galperin, A. Nitzan and I. Benjamin, *J. Phys. Chem. A* 106 (2002) 10790

[160] G. Binnig and H. Rohrer, *Surf. Sci.* 126 (1983) 236

[161] C. W. J. Beennakker and H. van Houten, textitSol. Stat. Phys. 44 (1991) 1

[162] R. Landauer, *IBM J. Res. Dev.* 1 (1957) 223

[163] B. J. van Wees, H. van Houten, C. W. J. Beenakker, J. G. Williamsan and C. T. Foxon, *Phys. Rev. Lett.* 60 (1988) 848

[164] A. G. M. Jansen, A. P. van Gelder and P. Wyder, *J. Phys. C* 13 (1980) 6073

[165] R. Landauer, *Z. Phys. B* 68 (1987) 217

[166] N. D. Lang, *Phys. Rev. B* 36 (1987) 8173

[167] P. W. Atkins, *Physikalische Chemie*, Weinheim, 1990

BIBLIOGRAPHY

[168] Y. Lilach, M. J. Iedema and J. P. Cowin, *Phys. Rev. Lett.* 98 (2007) 016105

[169] O. M. Magnussen, J. Hageböck, J, Hotlos and R. J. Behm, *Faraday Discuss.* 94 (1992) 329

[170] G. J. Edens, X. Gao and M. J. Weaver, *J. Electroanal. Chem.* 375 (1994) 357

[171] Z. Shi, j. Lipkowski, M. Gamboa, P. Zelenay and A. Wieckowski, *J. Electroanal. Chem.* 366 (1994) 317

[172] A. Cuesta, M. Kleinert and D. M. Kolb, *Phys. Chem. Chem. Phys.* 2 (2000) 5684

[173] R. M. Feenstra, J. A. Stroscio and A. P. Fein, *Surf. Sci.* 181 (1987) 295

[174] J. A. Stroscio, R. M. Feenstra and A. P. Fein, *Phys. Rev. Lett.* 57 (1986) 2579

[175] N. D. Lang, *Phys. Rev. B* 34 (1986) 5947

[176] K. S. Nakayama, T. Sugano, K. Ohmori. A. W. Signor and J. H. Weaver, *Surf. Sci.* 600 (2006) 716

Selected publications

- D.M. Kolb, F.C. Simeone,
  Electrochemical nanostructuring with an STM: a status report,
  Electrochimica Acta 50 (2005) 29089.

- D.M. Kolb, F.C. Simeone,
  Nanostructure formation at the solid/liquid interface,
  Current Opinion in Solid State and Materials Science 9 (2005) 91.

- F.C. Simeone, D.M. Kolb. S. Venkatachalam, T. Jacob,
  The Au(111)/Electrolyte Interface: A Tunnel-Spectroscopic and DFT Investigation,
  Angewandte Chemie International Edition 46 (2007) 8903.

- F.C. Simeone, D.M. Kolb. S. Venkatachalam, T. Jacob,
  Die Au(111)-Elektrolyt-Grenzschicht: eine Tunnelspektroscopie- und DFT-Untersuchung,
  Angewandte Chemie 119 (2007) 9061.

- F.C. Simeone, D.M. Kolb. S. Venkatachalam, T. Jacob,
  Tunneling behavior of electrified interfaces,
  Surface Science 602 (2008) 1401.

- P. Kaghazchi, F.C. Simeone, K.A. Soliman, L.A. Kibler, T. Jacob,
  Bridging the gap between nanoparticles and single crystal surfaces,
  Faraday Discussions, 140 (2009) 69.

- K.A. Soliman, F.C. Simeone, L. Kibler,
  Electrochemical behavior of Ir(210),
  Electrochemistry Communications 11 (2009) 31

- F.C. Simeone, C. Albonetti, M. Cavallini,
  Progress in Micro- and Nanopatterning via Electrochemical Lithography,
  Journal of Physical Chemistry C 113 (2009) 18987

Aknowledgements

First of all, I would like to thank Prof. Kolb. His human kindness and his scientific competence have been two fundamentals elements of my Ph.D.

Thanks to my colleagues Ludwig Kibler, Milla Manolova and Kahled Soliman for their trust and respect.

Special thanks to Ute Thomas, Constanze Duvigneau and Reinhard Liske...the other side of doing science!

Many thanks to Timo Jacob and Sudha Ventakachalam for having considered the dirty work of an experimentalist.

Thanks to Prof. Jürgen Heinze and Elmar Laubender of the University of Freiburg for their important cooperation. I can't forget Seda, Norman, Valentin, Florencia. Thanks to all of you: I felt part of the group.

The all staff and the members of the SFB569 are aknowledged for granting the project of my Ph.D. and for giving me the oppurtunity to work in a highly stimulating scientific network.

Die VDM Verlagsservicegesellschaft sucht für wissenschaftliche Verlage abgeschlossene und herausragende

Dissertationen, Habilitationen, Diplomarbeiten, Master Theses, Magisterarbeiten usw.

für die kostenlose Publikation als Fachbuch.

Sie verfügen über eine Arbeit, die hohen inhaltlichen und formalen Ansprüchen genügt, und haben Interesse an einer honorarvergüteten Publikation?

Dann senden Sie bitte erste Informationen über sich und Ihre Arbeit per Email an *info@vdm-vsg.de*.

Sie erhalten kurzfristig unser Feedback!

VDM Verlagsservicegesellschaft mbH
Dudweiler Landstr. 99   Telefon +49 681 3720 174
D - 66123 Saarbrücken   Fax   +49 681 3720 1749
www.vdm-vsg.de

Die VDM Verlagsservicegesellschaft mbH vertritt

Printed by Books on Demand GmbH, Norderstedt / Germany